Practical Power-Control Techniques

Irving M. Gottlieb

Howard W. Sams & Co.
A Division of Macmillan, Inc.
4300 West 62nd Street, Indianapolis, IN 46268 USA

© 1987 by Irving M. Gottlieb

FIRST EDITION
FIRST PRINTING—1986

All rights reserved. No part of this book shall be reproduced, stored in a retrieval system, or transmitted by any means, electronic, mechanical, photocopying, recording, or otherwise, without written permission from the publisher. No patent liability is assumed with respect to the use of the information contained herein. While every precaution has been taken in the preparation of this book, the publisher assumes no responsibility for errors or omissions. Neither is any liability assumed for damages resulting from the use of the information contained herein.

International Standard Book Number: 0-672-22493-3
Library of Congress Catalog Card Number: 86-62439

Acquired by: *Greg Michael*
Edited by: *Louis Keglovits*
Designed by: *T. R. Emrick*
Illustrated by: *Diversified Systems, Indianapolis*
Cover Art by: *Linda Simmons*
Composition by: *Photo Comp Corp., Brownsburg, IN*

Printed in the United States of America

Contents

Preface vii

CHAPTER 1 *OVERVIEW OF POWER CONTROL* 1

Ordinary Devices and Traditional Power Control 3
Regulated Power Supplies 6
Discrete Devices in Linear Voltage Regulators 7
Replacing Discrete Circuitry with Op Amps 10
Simplification of the Linear Regulator 11
Motor Control with Phase-Controlled Thyristors 13
Stereo Amplifiers Configured from Discrete Transistors 16
Stereo Power Supplies 18
Polyphase Versus Single-Phase Power 19
Relationships in Three-Phase Power Systems 20
Format of the Three-Phase Power System 22
Three-Phase Format from an Analog Oscillator 25
Three-Phase Format from an RC Bridge Network 27
Digitally Generated Three-Phase Format 28
Digitally Generated Quadrature Format 30
Power in the RF Spectrum 31
RF Power from MOSFETs 35
Drawbacks in RF Power MOSFETs 37
Power Interface between Logic and Action 38
BIMOS Power Switch 39
Three-Terminal Voltage Regulator 41
Three-Terminal Adjustable Voltage Regulator 44
Using Diodes to Protect Three-Terminal Regulators 45
Easy Way to Zap the IC Voltage Regulator 46
Thermal Hints and Kinks 47
Wiring Hints for Three-Terminal Regulators 49

CHAPTER 2 *SOLID-STATE AMPLIFIERS* 53

Simple Intercoms with Audio Power ICs 54
Optoisolators as Interstage Coupling Elements. 56
Getting the Best from MOSFET Power 58
A TV Sound Channel IC 60
Dedicated IC Low-Cost Audio Systems 62
Two Stereo Channels on a Chip 64
Bridge Amplifiers 66
Audio Power ICs for Vehicular Use 68
Complementary-Symmetry Channel Amplifier 72
20-Watt Workhorse Audio Amplifier 73
40-Watt Complementary-Symmetry Stereo Amplifier 75
Audio Amplifier with IC-Driven Output Stage................... 78
120-Watt Quasi-Complementary-Symmetry Stereo Amplifier 80
Servo or Motor-Drive Amplifier with 120-Volt Output 84
The RCA HC2000H Op Amp 86
RF Amplifiers for Mobile Service 88
A Mobile 70-Watt, 50-MHz Amplifier 89
A Mobile 25-Watt, 90-MHz Amplifier 89
A Mobile 10-Watt, 175-MHz Amplifier 91
A Mobile 20-Watt, 470-MHz Amplifier 92
Hybrid Power-Boost Module for Amateur 2-Meter Band 94
A 300-Watt, 2–30 MHz MOSFET Linear Amplifier 97

CHAPTER 3 *REGULATED POWER SUPPLIES* 101

Simple, Useful Single Transistor Power-Supply Circuits 102
Single Transistor Voltage Regulators............................105
Two-Terminal Constant-Current Diode 107
Quick and Easy Polarity Inversion 108
Tracking Regulator Using Three Active Devices 109
Synchronous Rectification 111
Power MOSFETs in a Synchronous Rectifier Bridge 113
Power MOSFET in Electronically Controlled Rectifier 116
Voltage-Regulating Power Supply with Current Foldback.........121
100-Watt, 100-kHz Switching Supply...........................123
Tracking a Negative Voltage...................................126
A Unique Step-Down Converter 128
A Multiple-Output Switcher 129
Multiple-Output, 230-Watt, 50-kHz Regulated Supply 131
A BIMOS Bridge Inverter 135

CHAPTER 4 *CONTROL OF ELECTRIC MOTORS* 141

A Variable-Speed DC Motor Controller.........................142

Efficient Speed Control of Permanent-Magnet DC Motors 143
Control of DC Motor Speed with a PWM Module 145
Unique Bidirectional Servo Drive System 148
Full-Wave Phase Control of Motor with an SCR 151
Controlling AC and DC Motors with a Power Op Amp 155
SCR Control Circuits for Series and Universal Motors 157
Simple Circuit for Speed Control and Regulation 159
AC Tachometer Functon via Special Optocouplers 161
Logic-Actuated Switch Provides Isolation......................166
Electronic Switch for Starting Induction Motors 168
Triac Phase-Control Circuit for Speed Adjustment 169
Transistors as Motor-Control Devices 171
Motor Control with a Dual Power Op Amp 172
Full-Wave Motor Controller Using Antiparallel SCRs............174
Speed Control of Larger Induction Motors.....................177
Dedicated IC for Controlling Four-Phase Stepping Motors 185

CHAPTER 5 A VARIETY OF USEFUL APPLICATIONS 193

Remote Switching of AC-Powered Loads 194
Integrated Hall-Effect Device for Switching Power 198
A Ruggedized Solid-State Relay 201
Solid-State Relays for Switching DC Loads.....................202
Interfacing a Microprocessor with an AC Load 204
Flashers...206
A Battery-Operated SCR Flasher 208
A Dedicated IC Suitable for Flasher Systems 209
Incandescent-Lamp Flasher Using a Thyristor 210
Flasher Using a PUT Oscillator...............................213
Stabilization of Laser Diode Optical Output Power..............215
An Electronic Siren 216
Power Op Amp Piezoelectric Buzzer Alarm System..............218
Power MOSFETs in AM Transmitters 219
Modulation Scheme for Power MOSFET RF Output Amplifiers ...221
Voltage-Regulator IC as an Amplitude Modulator................222
Power MOSFET in Horizontal-Sweep Circuits...................224
A Two-Transformer 100-Watt Ultrasonic Inverter 227
Capacitor-Discharge Ignition System..........................228
Simple 12-Volt Battery Charger...............................234
Burst Modulation Proportional Control of Heating System........235
A PWM Audio Amplifier 238

INDEX 243

Preface

The wonderful world of solid-state power control began with the germanium power transistor and the silicon controlled rectifier. These devices quickly made extensive inroads into power applications that had previously depended upon vacuum tubes, thyratrons, magnetic amplifiers, and electromagnetic relays. This is hardly suprising, for a number of compelling advantages were readily realized: smaller and lighter packages, relative freedom from mechanical and other wear mechanisms, more manageable and more convenient operating voltages, extended frequency response, and enhanced reliability, all at lower cost.

Because of current leakage and temperature dependency, germanium transistors gave way to silicon types; these, in turn, have undergone continuous progress in power, voltage, current and frequency capabilities, as well as in other parameters. SCRs, too, have evolved considerably over their earlier versions. Originally relegated to 60-Hz service, these devices developed to the stage wherein kilowatts of power could be controlled at many kilohertz. Also, the triac entered the scene, enabling full-wave control of AC power. From the mid-1960s to the early-1970s, this essentially remained the state of the art of power-control devices, and it appeared to some that this technology had attained maturity.

Beginning in the mid-1970s, however, and gathering steam during the early 1980s, a new trend in power devices and control systems became identifiable. For even though the bipolar power transistor had been marvelously improved, other devices began to assume prominence. Notably, the MOSFET shed its former flea-power status, becoming a heavyweight contender in power control. In somewhat similar fashion, the Darlington device invaded power levels and frequency domains previously thought unattainable. With these developments came the awareness that sophisticated ICs could greatly simplify implementation of the complex logic and driver circuitry often needed in conjunction with power devices. This, a notable achievement in itself, led also to the fabrication of logic, driver, and power stage on a single chip, or within a module, and probably suggests the nature of future progress in power-control techniques.

Such "smart" power devices may use a number of constructions and semiconductor formats. There are power ICs, MOS/bipolar integrations, power-device/control-logic combinations, hybrid modules uniting the best in power devices and analog or digital ICs, thyristors with self-contained trigger logic, optoelectronic devices with respectable power capability, and other implementations intended to reduce interconnections and number of parts.

The ensuing discussions and application examples should prove rewarding to the practitioner motivated to keep up with present and anticipated trends in the electronic control of power. In the interest of balance, respectable space has been allotted to many basic techniques, which, because of their time-proven usefulness, are unlikely soon to suffer extinction. The editorial format caters to those who like to build and experiment with their own circuits. Thus, the reader will find the presentations replete with component values, parts lists, and practical guidance.

IRVING M. GOTTLIEB

CHAPTER 1

Overview of Power Control

A new era is developing in the electronic control of power, and although its inception has been rather quiet, its implications are assuming considerable importance. What has happened is that continuous improvements to ordinary power devices have accumulated to the extent that sophisticated devices with previously unthinkable features have become available. Although former devices and circuit techniques are not all threatened with overnight obsolescence, many more options are becoming available to attain various goals in power control.

It is not always easy to identify the demarcation between "ordinary" power devices, and those which might do justice to such labels as sophisticated, exotic, or smart. For the most part, however, most bipolar power transistors of the types that have faithfully performed in myriads of power circuits, such as in regulators, amplifiers, and motor control can be said to be ordinary, or traditional devices. Admittedly, some bipolar power transistors with inordinate switching speed, frequency capability, or current or voltage ratings are certainly not garden-variety devices. Perhaps a reasonable test is to judge in terms of performance generally achieved five or ten years ago. If there has been a definite advance in certain parameters of performance, the device might well qualify as an extraordinary one, i.e., one that is of special interest for modern designs, and also as a possible forerunner of future development.

Our primary interest will be with power-control techniques making use of the newer and "smarter" power devices. This does not necessarily imply that the device must be integrated or associated with a microprocessor. Generally, however, there will be one or more plus features over devices commonly used a few years ago. For example, the modern power Darlington produces power gains not readily realized in the elemental power transistor. And, power MOSFETs require so little drive that they certainly merit classification as a modern power device destined for greatly expanded application in future designs of control circuits and systems. Then too, power op amps and power ICs can do things beyond the capabilities of ordinary power transistors, unless assisted by auxiliary devices and circuits. Even more suggestive of future trends are the now-developing families of dedicated devices and those with self-contained logic or drive circuitry.

The ultimate development of such smart power devices is a complete system or subsystem within a module. Fig. 1-1 depicts the basic idea—instead of a collection of discrete devices, stages, and associated passive components, all or most of the power-control function is accomplished within a single package. Not shown are internal protection against overload, transients, and temperature rise. This, of course is an idealized concept. There are thermal, electrical, fabrication, and cost factors which mitigate against universal realization of such an elegant format. Moreover, some tried-and-proven power control techniques using very ordinary power devices have been and still are satisfactory in practical applications. All things considered, the appreciation and mastery of the more modern power-control techniques will be best served by a general overview of the power devices and circuit techniques that have worked well in the recent past, and continue to do so in present power-control technology.

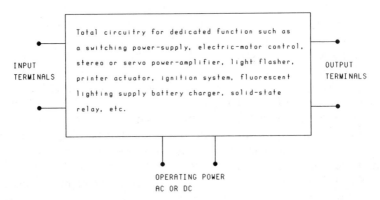

Fig. 1-1. The ultimate power device—a system or subsystem on a chip or within a module.

ORDINARY DEVICES AND TRADITIONAL POWER CONTROL

Although our objective is to investigate solid-state power devices with features, sophistications, and smarts that make them examples of and candidates for tomorrow's technology, it should prove meaningful to first review yesterday's state of the art. This is not a mere academic diversion from our alluded to goal; rather, it makes practical sense for several good reasons.

First, the trend in power electronics is to develop towards black boxes in which the ultimate format is a package with input and output terminals. Thus, no matter how complex the internally performed function is, properly processed power is delivered to the load with very little demand for understanding on the part of the user. From one vantage point, this is a desirable accomplishment—it certainly frees the user from brain-racking analysis, breadboarding and experimentation, and considerable time and expense. However, engineers, technicians, servicemen, and hobbyists tend to lose their skills and know-how unless they acquire some awareness of the internal circuitry of these devices.

The easy way to gain such knowledge is to consider the way the same overall function was, and to a large measure, continues to be done with discrete devices, or at least, with combinations of discrete devices and various ICs, such as op amps, logic circuits, voltage regulators, etc. After all, the designer of IC subsystems proceeds by duplicating or simulating individual circuit operations that previously were carried out in individual stages employing discrete devices. By monolithic integration or hybrid packaging, he then combines a large number of circuit functions into a small space. Whoever subsequently works with such an IC will not likely become cognizant of the intricacies of the device, but will, nonetheless, possess considerable insight if there is an appreciation of the manner in which the overall operation could be performed by brute-force use of discrete devices on a breadboard or PC board.

In this connection, many practitioners like to work with discrete devices during the initial phase of translating concept to hardware. Not only does this lead to a better feel for what will ultimately be achieved with packaged subsystems, but there is often greater freedom to measure, experiment, and evaluate. Indeed, in instances where there is little compulsion for optimizing efficiency, performance, or space, or where thermal considerations are very severe, it is, and will continue to be acceptable to implement systems largely, or entirely with discrete devices.

Second, many sophisticated power devices are better classified as discrete components rather than as subsystems, power ICs, or dedicated devices. Thus power MOSFET or power Darlington devices are not predesigned circuits; they fit right in with traditional ways of implementing solid-state power control. At the same time, these devices manifest capabilities grossly unrealizable a few years ago. Although these devices are not smart in their basic form, they appear to be destined for much future utilization in power control. By reviewing circuits and practices that have prevailed since the early stages of solid-state power, we will not be dealing with outdated technology if the earlier devices are replaced with newer ones. Here then is another reason why it is relevant to review "traditional" implementations of power devices. Interestingly, the symbology of schematic diagrams can be deceptive as to the true state of the art. An example is shown in Fig. 1-2, where a liberal interpretation in the drafting department may allow a common symbol to represent three different power devices.

Third, although black-box subsystems and superchips are the wave of the future for numerous applications, it is not the intent of this book to infer that all power control is destined to be accomplished via single smart modules. There is a compelling reason why it can be safely predicted that discrete devices will enjoy parallel development as technology progresses. This is because *thermal* limitations ultimately assert themselves as attempts are made to control more load power. It is remarkable that up to several hundred watts can be controlled by power

(A) NPN bipolar transistor.

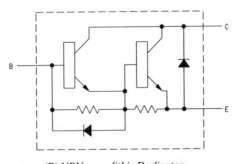

(B) NPN monolithic Darlington.

Fig. 1-2. A transistor is a transistor

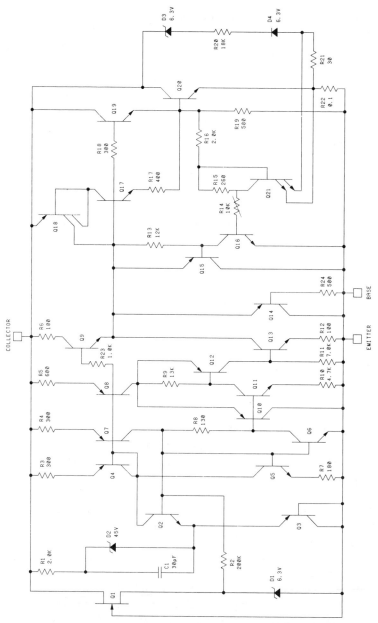

(C) Synthesized NPN power transistor. Other power ICs and power op amps could also be represented by the transistor symbol in (A).

—not necessarily.

op amps and other ICs with power-output stages. But, even though much of the discussion in this book will be focused on these devices, it remains hard fact that higher powers must be handled by discrete devices, just as in the past. Of course, modern technology allows us to incorporate the "brains" of high-power control in ICs and electronic subsystems. And, modern technology also provides discrete power devices which greatly outperform the earlier bipolar transistors.

Fourth, when one considers a number of aspects involved in manufacturing, it becomes reasonable to anticipate that many power-control systems are destined to be a mixture of techniques, involving integrated or hybrid subsystems, dedicated ICs, and discrete devices. Chief among these production factors are cost, availability, and thermal considerations.

Summing up, practical power control will make use of power devices with self-contained auxiliary functions, but much of traditional stage-by-stage implementation is likely to be retained. An understanding of how basic circuit functions are carried out with discrete devices will provide meaningful insight into power-control systems, whether they comprise a combination of individual stages, or at the other extreme, a single device predesigned to carry out the desired operation.

REGULATED POWER SUPPLIES

Regulated power supplies have become important subsystems in a wide variety of electronics equipment. One reason is that a predictable and stabilized level of DC operating voltage greatly enhances the reliability, uniformity, and manufacturing reproducibility of electronic circuitry. Moreover, with the tendency to use direct coupling between stages, tolerance to changes in operating voltage generally is not as good as in AC coupled stages. The trend towards direct coupling has already been in evidence in discrete-device circuits where simplicity and reduced parts count results. In monolithic integration, however, direct coupling is almost mandatory in the fabrication process. Thus, virtually all ICs and sophisticated power devices utilizing more than a single element employ direct coupling between internal stages.

An even more compelling reason, although not always a fully appreciated one, is that a voltage-regulated power supply exhibits very low AC impedance to the circuitry being powered. This keeps AC and signal currents out of the power source, enabling it to supply DC operating power, but isolating it from otherwise becoming part of the powered circuit. This is extremely important if the powered circuit is to operate predictably and as intended. This feature of the voltage-regulated source makes it readily feasible to operate many cascaded stages in amplifiers and subsystems without feedback or interaction between the stages.

This attribute of regulated supplies is valuable to the extent that many circuits would operate satisfactorily from a regulated supply providing only the low output impedance feature and no stabilization of the DC voltage level at all. Indeed, such arrangements have been used in TV sets and in hi-fi amplifiers. It helps visualization to look upon the regulated supply's low AC impedance as deriving from a giant simulated capacitor—one much larger than could be practically implemented with an actual physical capacitor.

Regulated power supplies can be used to regulate current instead of voltage. In this mode, they have been used for linearizing ramp waveforms and for simulating high-impedance loads in amplifiers; they are particularly useful in differential amplifiers. In power applications, the current-regulated supply often looms up as the most appropriate power source for motors and solenoids inasmuch as the torque of such electromechanical devices is a function of current, that is, of ampere turns. Electrochemical processes, such as battery charging and electroplating, are often best operated from current-regulated sources. And, welders tend to be essentially current-regulated sources.

Sophisticated power devices have been improving all aspects of regulator design, performance, packaging parameters, cost, and friendliness towards powered equipment. Although the underlying theory remains about the same, modern-device regulators show much progress over traditional designs—those based upon many discrete devices.

DISCRETE DEVICES IN LINEAR VOLTAGE REGULATORS

The examples of linear (nonswitching) voltage regulators shown in Fig. 1-3 depict a sequence from the simplest to fairly sophisticated. The sequence from Fig. 1-3A to 1-3D also represents the evolution of this basic type of regulator. Of particular interest are the configurations of Fig. 1-3C and D. The regulator in Fig. 1-3C is about the simplest arrangement of discrete devices that yields all-around satisfactory performance for a large number of applications, but the more primitive regulators of Figs. 1-3A and B have been useful where requirements have been relatively relaxed and where cost has assumed major consideration. The circuit in Fig. 1-3D may be considered a "beefed up" version of the circuit in C and incorporates the following basic advantages over it.

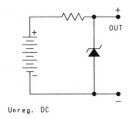

(A) Zener diode DC supply.

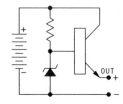

(B) Emitter follower, or "amplified" zener circuit.

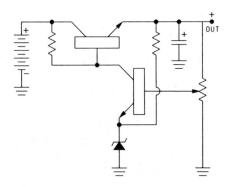

(C) Simple "workhorse" regulated supply.

Fig. 1-3. Representative voltage regulators

- It has a differential error amplifier.
- It uses a dedicated voltage reference, rather than a plain zener diode.
- It has an extra stage of error-signal amplification.
- It has a Darlington output arrangement for the series regulating element.
- It has an output current-limiting provision.

The performance of regulator in Fig. 1-3D greatly exceeds that readily attainable from the more basic arrangement of that in C. It is capable of tight line and load regulation, very low dynamic (AC) impedance, low temperature coefficient, and has built-in protection for the series-pass device. Nonetheless, the very popularization of voltage regulators similar to that of the circuit in D provided a strong incentive towards further progress in the state of the art. Let us see why this has been so.

Even a casual inspection of the regulator of Fig. 1-3D shows that it has become relatively complex in configuration and parts count; certainly the inclusion of additional discrete stages would begin to make it awkward and unwieldly. Unfortunately, considerably more add-on circuitry was often needed. This is because the regulator is still a "barebones" arrangement, for the systems designer often required extended

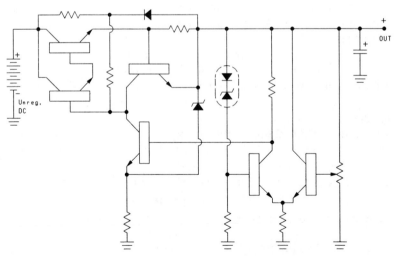

(D) A more sophisticated voltage-regulated supply.
using discrete elements.

performance capabilities, such as even closer regulating behavior, thermal shutdown, overvoltage protection for loads, SOA (safe operating area) protection of the series-pass device, remote sensing, zero-adjust capability for output voltage, optional or automatic current regulation, transient protection, load-current sinking, easy paralleling capability, digital programming, dissipation control, negative or other additional output voltages, etc.

The bottom line here is that a large number of discrete transistor stages tends to bear adversely on reliability, reproducibility, stability, and cost. Although the concepts residing in the regulator of Fig. 1-3D remain valid, a better way has been found for practical implementation. This entails the use of IC modules to handle most low-level circuit functions.

REPLACING DISCRETE CIRCUITRY WITH OP AMPS

Before the advent of dedicated IC regulator circuits such as the LM723, considerable reduction in parts count together with upgraded performance was achieved by the use of op amps. Most commonly, these substituted for two and often three discrete transistors which would otherwise be required for the error-amplifier circuitry. Commonly, the 741 garden variety op amp has been found to work satisfactorily in this function. Fast acting op amps of the voltage-comparator type have also served this purpose, although their use is of greater advantage in switching, rather than in linear regulators.

The basic circuit configuration of a popular form of this type of regulator is shown in Fig. 1-4. The plus and minus signs on the op amp refer respectively to the noninvert and invert input terminals. Note that the op amp provides a differential input stage, whereas a single transistor does not. Aside from compact packaging, simplification, and cost saving, the salient feature of the op amp is the very high amplification that can be imparted to the error signal without oscillation or instability. In this regard, it will be noted that the op amp in Fig. 1-4 is run wide open—there is no gain-reducing negative feedback loop, as is usually the case in instrumentation and audio amplifiers.

Op amps are very versatile devices and there are many different ones available. They differ in gain, frequency response, operating voltage, etc., but most incorporate from one to several dozen active devices. Although they tend to have three main terminals (a differential input pair and an output terminal), their internal circuitries can differ markedly.

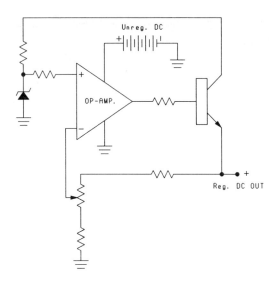

Fig. 1-4. Basic setup for linear voltage regulators using op amps.

Yet, their operating characteristics are precisely specified and they have been designed to produce clean operation. Thus, they are at the outset easier to implement than discrete transistor circuitry. Indeed, a regulator with a high gain error amplifier using discretes often requires a lot of cut-an-try experimentation to obtain stable operation over a range of load and line conditions. Even worse, as both hobbyists and manufacturers have sadly learned, the attainment of stable high-gain operation from cascaded transistor stages does not ensure like results from the next batch of "similar" transistors.

Of course, once these factors were fully realized, the next step was to devise dedicated ICs especially intended for regulators. These proved even better than op amps, for they incorporated precise voltage-reference sources, as well as other provisions that reduced the external parts count of the overall regulator. One of the early ICs of this type was the 723 regulator. This became a universal workhorse in power supply design and the end of its usefulness is not yet in sight.

SIMPLIFICATION OF THE LINEAR REGULATOR

Fig. 1-5 shows a linear voltage regulator which can be implemented as the approximate electrical equivalent of the all-discrete regulator of

Fig. 1-1D. The LM723 voltage regulator IC handles low-level circuitry functions, and the load power is handled by the external resistor. The simplification is dramatically obvious; it is not surprising that the LM723 and similar ICs have long been used as the "brains" of such voltage regulators. Fig. 1-6 depicts the circuit of the LM723. Note that there are 16 active devices. It is easily appreciated that the brute-force construction of such a circuit from discrete devices would be a formidable, awkward, and costly task.

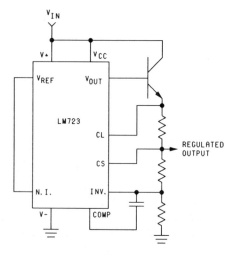

Fig. 1-5. Example of simplification of linear regulator via the use of an IC.

As might be expected, the regulator of Fig. 1-5 can outperform the discrete model of Fig. 1-1D. This is why the two circuits were declared to be "approximately" equivalent. But, there is even more than meets the eye, for the LM723 dedicated IC is endowed with versatility and flexibility not readily attained in all-discrete configurations, or even in layouts in which some discrete devices are eliminated by using nondedicated ICs, such as op amps and voltage comparators. With relative ease, the LM723 can be connected to a small number of external components to provide the functions of positive- or negative-voltage regulator, current regulator, shunt regulator, floating regulator, high-voltage regulator, variable-voltage regulator, foldback, or current-limiting regulator, and several different modes of switching regulators. It can be used in remote (load) sensing applications, and can be controlled from logic signals. In all of these techniques, a small capacitor suffices for frequency compensation, rendering the regulator unconditionally stable, regardless of the nature of the load.

Comment is in order with regard to the use of this versatile IC in switching regulators. This was *not* its original intent, inasmuch as switching regulators were rarely used when the LM723 made its market debut. Of course, one of the reasons switchers initially had a reputation for unreliability and generally bad performance was because of implementations with all-discrete circuitries. Because of the high-frequency harmonic energy in switching suppliers, these circuits cannot be constructed in the manner of other applications dealing with DC, low-frequency AC, and audio currents. But the resort to miniaturization and RF techniques is difficult and expensive, especially when a large number of discrete devices are involved. It was fortuitous that the LM723 proved a very capable performer in switching regulator circuits. Although dedicated ICs for switching regulators eventually enjoyed prime consideration, it remains remarkable that very satisfactory switching regulators can be put together with the LM723, and a few external components. As with linear regulators, very high current and power levels can be attained through the use of appropriate external load-handling devices.

MOTOR CONTROL WITH PHASE-CONTROLLED THYRISTORS

The motor-control circuit shown in Fig. 1-7 is essentially the same as that used in light dimmers. In this particular example, an SCR is used and it is triggered by the most elemental device available for the purpose—a small neon lamp. This circuit is very basic; it lacks line filtering and is somewhat plagued by erratic performance, particularly if too great a control range is attempted. Moreover, it doesn't make optimum use of the motor's torque characteristics because of the half-wave rectification provided by the SCR.

Although lacking in sophistication, this basic scheme is representative of widely used circuits using phase-controlled thyristors. The simplicity of this circuit is compelling and it is easy to add on various modifications. The universal motor is very similar to a series-type DC motor, but its design is optimized to enable it to run directly from AC with minimal heating or sparking. The basic concept of speed control derives from control of the timing of the gate triggering voltage for the SCR. A nice aspect of this method is that no commutating provision is needed for the SCR. This has probably been a major reason for the popularity of phase-control circuits using thyristors. Thyristors, by virtue of their operating efficiency and power capability, are almost ideally suited for motor control.

This method is not suitable for AC induction motors. This has nothing to do with the phase-controlled thyristor, but it is because induction motors have very little voltage sensitivity—their speed is far more a function of frequency than it is of impressed voltage. But by the same reasoning, this circuit can be used for DC motors under appropriate circumstances. This follows from the fact that the controlled current is unidirectional and is, accordingly, a form of DC.

Despite its simplicity, this basic circuit is not power limited by the control device. On the other hand, motor-control circuits using bipolar transistors have had difficulty in coping with the current demands as well as implementation and operating costs of large motors. The built-in advantage of thyristors has been that they switch on and off very rapidly and develop small dissipative loss while conducting. As will be seen in a forthcoming chapter, this privileged position is now being successfully challenged by other devices that can now control medium-sized motors.

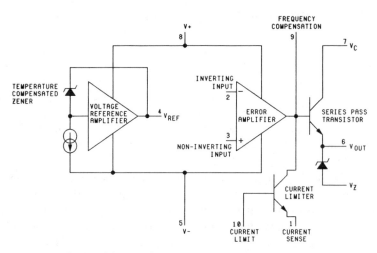

*Pin numbers refer to metal can package.

(A) Functional block diagram.

Fig. 1-6. An inside look at the 723 IC

14 / POWER-CONTROL TECHNIQUES

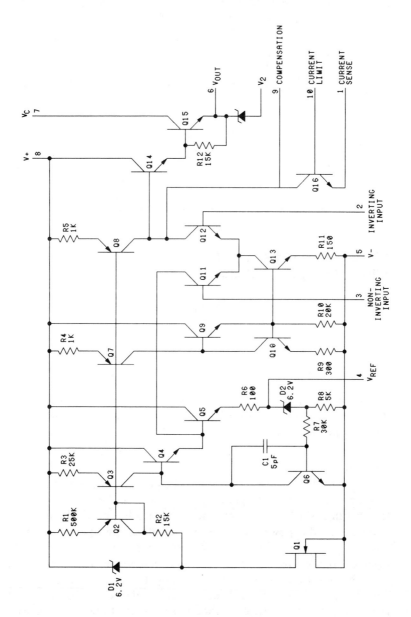

(B) Equivalent schematic circuit.

voltage regulator. (Courtesy National Semiconductor Corp.)

OVERVIEW OF POWER CONTROL / 15

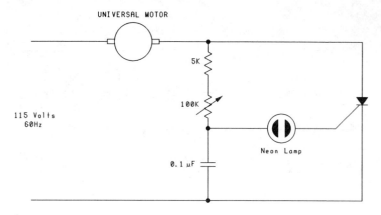

Fig. 1-7. Simple speed control for small universal motors.

STEREO AMPLIFIERS CONFIGURED FROM DISCRETE TRANSISTORS

Fig. 1-8 is a representative example of the solid-state stereo amplifiers that have been produced during the '70s and the early '80s. As can be seen, a large number of transistors are used in a rather complex circuit. The fact that DC coupling is used throughout is quite an accomplishment considering the wide tolerances exhibited by garden-variety transistors. Actually, this particular circuit incorporates the modern feature of complementary-symmetry power Darlingtons in the output stage. This permits direct connection to the voice coil of the speaker without any coupling capacitor.

Earlier versions of this basic design approach used single device NPN and PNP transistors in the output stage rather than the Darlingtons shown. And before this, the so-called quasi-complementary symmetry circuit prevailed; this output circuit used two NPN power transistors because silicon PNP types were not readily available yet. This required that one power transistor of the output pair operate as a common-emitter stage, whereas its mate was connected as an emitter follower. A phase inverter in the driver circuit made the output stage operate in push-pull fashion. At first thought it would seem that such a patchwork arrangement would be grossly imbalanced. This was not the case, however; the relatively low load impedance of the speaker caused both transistors to behave quite similarly. Single power-supply amplifiers needed a large electrolytic capacitor to couple the audio power to the

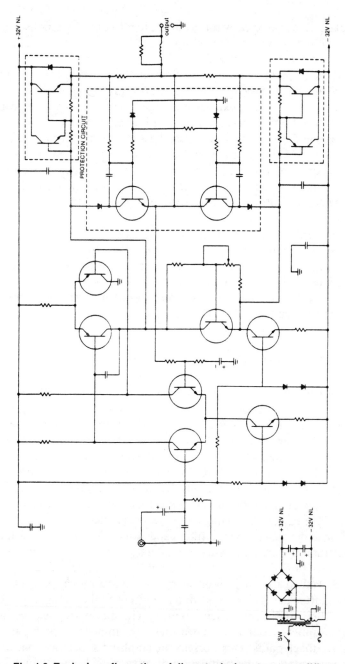

Fig. 1-8. Typical configuration of discrete device stereo amplifier.

speaker. This caused early rolloff of the low-frequency response, but its main objection was its change of characteristics over time and, often, its tendency towards failure.

Of course, a large output transformer would have been preferable, except for the expense thereby entailed. A transformer would not necessarily do better at low frequencies than a large output coupling capacitor, but would likely outlast virtually all other components.

Discrete-device audio circuits can be a serviceman's headache, for if anything goes wrong in any of the stages, a chain reaction often kills several, or more, transistors. Yet, once everything is in order, reasonably good reliability has been the track record of such circuits. Moreover, these basic circuits have been successfully designed for a wide range of output levels.

The performance of the quasi-complementary amplifier has been considerably enhanced by the fact that the semiconductor firms have been making dedicated power transistors for this purpose. These minimize crossover distortion and maximize linearity and balance. Practically both the complementary symmetry and the quasi-complementary symmetry circuits are capable of good results.

STEREO POWER SUPPLIES

The elementary power supply shown in conjunction with the amplifier of Fig. 1-8 represents a compromise dominated by cost considerations. Amplifier performance is largely dependent upon the large electrolytic filter capacitors in these simple DC supplies. Because of the cost factor, practical limits restrict the size of these capacitors; even worse, their capacity tends to decrease with time and their internal resistance increases. As a result, low frequency response, power availability, and distortion of the amplifier are not as good as they might be with giant filter capacitors having more ideal characteristics.

The dependencies on the filter capacitors arise from considerations other than the mere necessity to attenuate power supply hum. Despite the use of push-pull circuits in the amplifier, a residue of audio currents are forced to use the power supply as circuit return paths. (Driver stages, often being single-ended, rely entirely on the DC supply to complete their audio-frequency circuits.) The audio frequencies in the power supply encounter nonlinearities, high-impedance paths, and mutual feedback paths. This worsens the amplifier's specifications on frequency response, intermodulation distortion, and power capability at extremes of frequency response.

It has long been known that the answer to the problem incurred through the use of simple DC supplies is the voltage-regulated power supply. A quick way of appraising such a remedy is to view the regulated supply as the electronic equivalent of an extremely large filter capacitor. Actually, it is even better than this, for a giant physical capacitor would still offer higher reactance to lower frequencies. The electronic capacitor (the regulated supply) on the other hand, offers a low constant-impedance to frequencies right down to zero, or DC.

Because of developments in IC regulators and other device improvements, it has become economic to use voltage-regulated power supplies in high-performance stereo amplifiers, and this represents the future design trend. Other benefits also accrue, such as the reduction of background 60 and 120 Hz hum, and the reproducibility of amplifier control settings. Additionally, there tends to be better immunity to power-line transients. Finally, the voltage-regulated supply is much less susceptible to the adverse effects of time and temperature. Both linear and switching regulators have been successfully utilized, with the switching type progressively gaining in use, especially for higher-power amplifiers. Of course, care must be exercised in implementation that the switching supply does not inject noise or add other problems.

POLYPHASE VERSUS SINGLE-PHASE POWER

Polyphase power systems have a number of advantages over more simple single-phase systems. Indeed, it is not always true that simplicity is necessarily associated with the single-phase approach. The advantages are well known because of the long exploitation of the three-phase format in utility and industrial power systems. Because of shortcomings in power-handling capability, reliability, and cost, solid-state power techniques were slow to make their appearance in polyphase systems. That era is over. Bipolar power transistors, power MOSFETs, and thyristors have evolved to the point where respectable power levels can be implemented. At the same time, logic devices and techniques make waveform synthesis feasible with surprisingly few IC modules. All in all, polyphase power systems can be said to possess the following advantages over the single-phase format:

- Three-phase induction motors are superior to single-phase AC motors in nearly all respects.
- Two-phase motors are excellent for use in servo systems. They are cheaper and more maintenance free than are DC motors. And,

unlike single-phase induction motors, they are self-starting.
- For airborne and space applications, three-phase transformers provide weight reduction over single-phase units working at the same power level.
- Rectified polyphase voltage is easier to filter than the rectified voltage from a single-phase source.
- A three-phase inverter or power supply maintains power balance in a three-phase source. By contrast, single-phase inverters or power supplies contribute to the imbalance of power in a three-phase power source.
- Three-phase transformer windings can be connected so that much of the third-harmonic energy in a square or quasi-square wave is attenuated. Such electrical or electronic filtering reduces the need to rely upon bulky physical filters. This is of considerable importance in the operation of motors; harmonics can cause temperature rise and rough operation. The cancellation of the third harmonic also removes higher-order odd harmonics which often cause electrical interference with other systems.
- The manner of connecting the windings of three-phase transformers also provides the designer and user with additional flexibility in the selection of output voltage and available current.
- Logic-circuitry for generating the polyphase format can be readily manipulated to provide various performance features. These include dead-time intervals, stepped shapes to simulate sine waves, changing phase sequence for reversing induction motors, harmonic manipulation, frequency variation for speed control of induction and synchronous motors, and output-voltage control. Such logic circuitry also facilitates feedback applications for maintaining constant speed, torque, or horsepower.

RELATIONSHIPS IN THREE-PHASE POWER SYSTEMS

The basic source or load connections encountered in three-phase systems are shown in Fig. 1-9. In these illustrations, resistive loads are shown connected in symmetrical formats to receive balanced power from their three-phase sources. It is assumed that the three loads in each diagram are equal. Because of the phase relationships of the three source voltages, common-sense arithmetic leads to erroneous calculations. For example, in the delta connection of Fig. 1-9A the line current, I, is not three times the current that flows in each load element. Rather, the line current is 1.732 times an individual load current. This is a basic relationship and derives from the trigonometric relationships in three-phase systems. Similarly, in the Y connection of Fig. 1-9B, phase voltage or line voltage is 1.732 times the voltage across an individual load. The number, 1.732 is an ever-present factor in three-phase calculations; it is the square-root of 3, and is also the tangent function of 120 degrees.

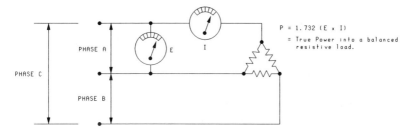

(A) Delta-connected loads receiving three-phase power.

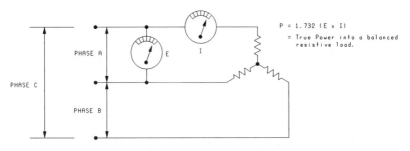

(B) Y-connected loads receiving three-phase power.

Fig. 1-9. Two basic connections used in three-phase circuits.

Note that the phase voltages in the delta connection are equal to the individual load voltages. Here is a case where one can state that the equality is obvious from inspection and not have to suffer subsequent embarrassment for ignoring some subtlety of mathematics. Reciprocally, we can see that the line current and the individual load currents in the Y connection are the same.

OVERVIEW OF POWER CONTROL / 21

From the above considerations, it turns out that total load power, P, in both delta and Y systems is given by P = 1.732 (E × I), with E and I as depicted in Fig. 1-9. The common mistake made by those not familiar with three-phase relationships is to suppose that P = 3 (E × I), a natural enough error, deriving from experience with single-phase systems.

If the loads have reactance as well as resistance, the equation for true power, P, becomes P = 1.732 (E × I)(PF) where PF is the power factor of the load impedances. One definition of power factor is that it is the ratio obtained by dividing true power by apparent power. Apparent power, when the loads are impedances, is the previous expression, P = 1.732 (E × I). True power can be obtained from wattmeter measurements. However, if the impedance, Z, and resistance, R, of the load elements are known, the wattmeter measurements are not needed. This is because PF = R/Z. Summarizing, True Power = 1.732 (E × I)(R/Z) for all load impedances in both delta and Y systems, where Z is the individual load impedance, and R is the individual load resistance.

Polyphase power can be likened to a multicylinder engine—there is a more continuous flow of power than in a one-cylinder (single-phase) energy source. Although this analogy may not be very meaningful for resistive loads, a polyphase motor does have a smoother rotational torque than a single-phase motor. This has been one of the salient features of three-phase motors in industry. Now, via power electronics, such motors can be smoothly controlled with relatively inexpensive equipment. This is particularly significant for electric vehicles and traction applications, for it has become feasible to convert either DC or single-phase power into the polyphase format (usually three phase) by solid-state power devices.

FORMAT OF THE THREE-PHASE POWER SYSTEM

Considerable insight into the nature and behavior of three-phase systems can be gained by a close scrutiny of the three 120-degree displaced sine waves shown in Fig. 1-10A. At any instant of time, the algebraic sum of the three waves is zero. This is important and is not obvious from casual inspection. A near proof of this relationship is as follows. Consider the situation when one of the waves is at its maximum value. At that time, it will be seen that the other two waves are both at 50% of their maximum values, but of the opposite polarity with respect to the first wave. Thus, the algebraic sum of the three waves at that time is zero. Again, consider the situation when one of the waves is of zero amplitude. At that time, it will be seen that the other two waves are at approximately 86% of their respective amplitudes, but of opposite polarity to each other. As before, the algebraic sum of the three waves is zero. It can be further shown that this relationship holds for any time between zero and three hundred and sixty degrees for any single wave. Or more simply stated, the relationship is true at any time. The vector representation shown in Fig. 1-10B provides additional insight into the three-phase format of sinusoidal voltages.

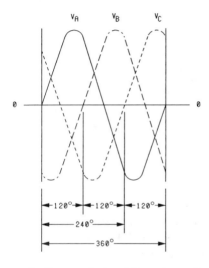

(A) Oscilloscope display of three-phase voltage.

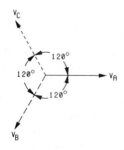

(B) Vector diagram representation of the sine waves of Fig. 1-10A.

Fig. 1-10. Format of voltages in a three-phase power system.

A practical consequence of this relationship is shown in Fig. 1-11A where the coils represent either the output windings of an alternator or the secondary windings of transformers. The coils, in other words, are sources of three-phase voltages. The closed configuration, known as the delta connection, superficially appears to be a short circuit. But if the delta were opened, as depicted in Fig. 1-11B, it would be found that zero voltage exists between terminals X and Y. This is in accordance with the logic developed in the preceding paragraph. Thus, it is possible to close the delta. But two things must be kept in mind.

(A) Delta connection for three 120-degree displaced voltages.

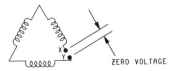

(B) In a proper delta system, no voltage exists between any of the coils at their corner connections.

Fig. 1-11. Alternator output windings or transformer secondaries connected in delta.

First, the delta can only be closed for sine waves, or for other waves having no third, sixth, or ninth harmonics (or higher integral orders of third-harmonic energy). Such harmonics do not sum up to zero as does the fundamental frequency. Therefore, the delta connection will constitute a short circuit to such harmonics. (It will be shown later how this can be advantageously used in certain situations.)

Second, three sources of 120-degree displaced voltages cannot be indiscriminately connected in the delta configuration. Attention must be given to the phasing of the sources. In practice, this means that two of the sources can be connected together without regard to which terminals are selected, but the remaining source can only be connected one way. If it is connected the wrong way, a short circuit will result.

It is well to be mindful that many electrical engineering texts deal primarily with sine wave polyphase systems. In contrast, electronic inverters generally make use of square, quasi-square, or stepped waveforms. What is valid practice for sine waves may or may not be valid for these nonsinusoidal waves.

THREE-PHASE FORMAT FROM AN ANALOG OSCILLATOR

A three-phase sine-wave format can be produced by cascaded amplifying devices in a symmetrical circuit configuration such as shown in Fig. 1-12. In essence, this amounts to a three-stage RC coupled amplifier in which the output is fed back to the input. Inasmuch as a 360-degree phase shift is needed to produce oscillation, it is apparent that such a scheme would not be oscillatory if only the phase-shift occurring in the op amps was effective. What is needed is an additional 180 degrees of phase shift. This is achieved via the RC coupling networks at the output of each stage. The overall circuit oscillates at that frequency at which the total RC phase shift is 180 degrees. This means that each RC network provides 60 degrees of phase shift. When one considers the total per-stage phase shift—that provided by an op amp in conjunction to that occurring in its output RC network—it turns out that 120 degrees of phase displacement exists between any two of the outputs. Thus, phase A is displaced 120 degrees from phase B in one direction and 120 degrees from phase C in the opposite direction. This amounts to a three-phase format of sine waves.

Note that as long as all R's are equal and all C's are equal, the three-phase format is preserved; only the frequency changes with a change in R or in C. Practical values of C will often fall between 0.1 and 1.0 μF over the 50-Hz to 1000-Hz frequency range. Then, R = X_c/Tan 60°, where X_c = $1/2\pi fC$, and the tangent function of 60 degrees is 1.732. By utilizing three ganged pots, a wide control range of three-phase frequencies may be had. This is very useful for speed control of three-phase induction or synchronous motors.

The op amps may be any of the numerous designs based upon the original 741. Later versions of this op amp have FET inputs and there are often other sophistications. The circuit is not demanding and actually many garden-variety op amps will prove satisfactory. In any event, the basic idea is to generate good sine waves. This may require lower values than the 150K feedback resistances shown. If square waves are needed, they are best obtained by signal processing, following the basic circuit of Fig. 1-12. In any event, buffer and power amplifiers will be required in each phase; square waves can be produced by allowing such amplifiers to be overdriven or saturated, or by means of triggered logic devices. Whatever amplifying or wave-shaping circuits are used, they should be identical for the three phases. Otherwise, the interphase balance is likely to be destroyed, with deleterious effects on both the inverter and the load.

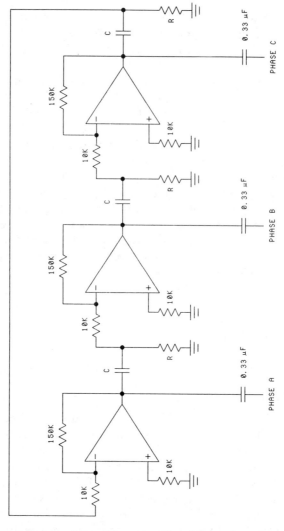

Fig. 1-12. Generalized circuit for generation of three-phase format by analog technique.

DERIVATION OF THREE-PHASE FORMAT FROM AN RC BRIDGE NETWORK

An RC bridge in conjunction with a center-tapped transformer can be utilized to produce a three-phase format of voltages if driven from a sine wave source. Such an arrangement is shown in Fig. 1-13. Because of tolerances in components and the difficulty of obtaining exact mathematically calculated values, some experimentation may be in order to produce three equal-amplitude voltages, spaced 120 degrees in phase. The component values depicted in the diagram will come close, however. This circuit is frequency sensitive and will not work with square waves because the harmonic frequencies will not be accorded proper phase shift or amplitude response. If square waves are needed from the three-phase format, appropriate squaring circuits can be introduced following the three-phase conversion.

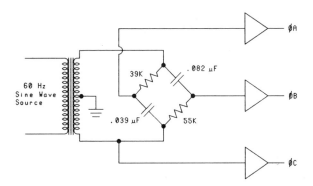

Fig. 1-13. Single phase to three phase conversion with an RC bridge.

It will be noted in Fig. 1-13 that one of the four RC junctions in the bridge is not used for providing an output voltage in the three-phase format. It is important that the unused junction be the one indicated; although the schematic diagram appears to be topographically symmetrical, only the selected three junctions will provide the required 120-degree phase displacements.

In the event a sine-wave frequency other than 60 Hz is involved, the capacitors in the bridge should be changed to offer the same reactance as the indicated ones do at 60 Hz. Thus, if the input frequency were 600 Hz, the capacitors would be changed to one-tenth of their 60-Hz values. The resistances in the bridge can be retained at their 60-Hz values for any frequency.

Minor changes in the amplitude of the phases can be made by varying the gain of the amplifiers. Reversed phase rotation can be accomplished by transposing any two amplifier-input connections to the bridge. Such phase rotation changes the direction of rotation of a three-phase induction or synchronous motor. It is also an important consideration when two or more three-phase sources are to be operated in parallel for increased power capability—all such sources must have the same sequence of phase rotation and this must be determined before the sources are paralleled. Otherwise, destructive short-circuit currents will flow. (The other requisites for paralleling three-phase sources of power is that all voltages should be balanced and equal, and all frequencies should be the same.)

DIGITALLY GENERATED THREE-PHASE FORMAT

A generally more useful way to generate the three-phase format than by analog circuits is by the utilization of digital logic. Not only is the square-wave output thereby produced usually desired for inverters, converters, and power supplies, but the precision of timed waves is not dependent upon passive component values. Most digital designs make use of a three flip-flop shift register arranged as a "twisted-ring" counter. This basic scheme is shown in Fig. 1-14. Here the flip-flops are JK binaries. When this ring counter is clocked at six times the desired frequency of the three-phase format, a set of square waves is generated which complies with the timing sequence of the three-phase format.

The simple setup of Fig. 1-14 has a shortcoming, however. Although the truth table is essentially correct, it does not tell the whole story. In addition to the six indicated logic-level combinations, there is also the possibility of the counter locking up in either 000 or 111. Such lockup tends to occur when the counter is switched on (when DC operating power is first applied), or when a noise transient appears during ordinary operation. What is needed is additional logic to *prevent* the occurrence of these undesired states.

A more practical form of the digital three-phase generator is shown in Fig. 1-15. The additional NOR and OR gates prevent the counter from deviating from the truth table of Fig. 1-14. Otherwise, the performance is essentially the same as in the basic setup. Although there is a large variety of these three-phase generators using various types of flip-flops and a variety of logic gates for inhibiting the undesired counter state, most operate very similarly to the circuit of Fig. 1-15.

If a crystal-controlled clock is employed, excellent precision in both frequency and phase balance is possible. For some purposes, the clock can be synchronized to the 60-Hz utility source. The digital method of

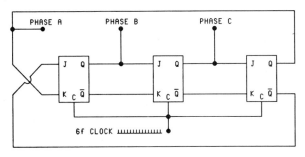

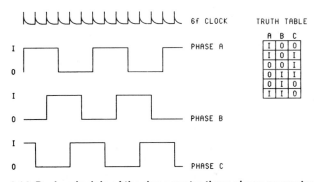

Fig. 1-14. Basic principle of the ring counter three-phase generator.

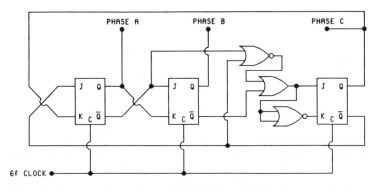

Fig. 1-15. Same basic circuit as Fig. 1-14 with addition of logic gates to prevent operation in undesired modes.

three-phase generation merits consideration even if the ultimate output must be sinusoidal, for such conversion can be achieved via a simple LC low-pass filter, or by an active bandpass filter.

DIGITALLY GENERATED QUADRATURE FORMAT

Two-phase induction motors are very useful in AC positioning and servo systems. Such motors are mechanically rugged and need little maintenance. Much like three-phase motors, they are inherently self-starting, and electrically reversible. The required voltage format for these motors is two equal-amplitude waves with the phase displacement between them equal to 90 degrees. At first consideration, such a format may appear to be unsymmetrical. Nonetheless, such a quadrature-phase format results in very smooth and efficient motor operation.

The traditional way of producing the 90-degree phase difference between the two motor windings is to insert a capacitor in series with one winding. See Fig. 1-16. Although satisfactory for many applications, exact quadrature phase displacement is not attained, and the electrolytic capacitors often used tend to change their capacity and increase their leakage current and effective series resistance with temperature and age. Moreover, if the speed of the motor is to be varied by changing the applied frequency, different capacitors must be switched in and out. This is both awkward and costly. The motor itself will usually operate well over a wide speed range providing the quadrature relationship is maintained in the two-phase supply. (For best results, it may also be necessary to increase voltage with frequency in order to maintain a constant motor current.)

An elegant way to generate a precise two-phase format for the motor is with a digital circuit such as is shown in Fig. 1-17. The two D binaries produce the accompanying waveforms of voltage. Note the requirement for 4f clock pulses. The active filters, either low pass or bandpass, convert the square waves to sine waves. Depending upon the motor used, these filters may not be necessary. Often, the motor inductance suffices to make motor current a close-enough approximation to a sine wave so that there is no undue temperature rise or torque disturbance from the residual harmonic energy. If the filters are used and a wide frequency range is employed for speed control, it is much easier to change the RC networks in such active filters than it is to change the large capacitor in the conventional phase-shifting technique.

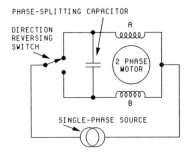

(A) Circuit.

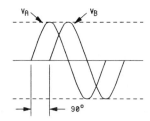

(B) Quadrature voltage format applied to motor.

Fig. 1-16. Conventional way of supplying a motor with a two-phase format.

POWER IN THE RF SPECTRUM

At one time the use of bipolar transistors to develop appreciable power at high frequencies was in the realm of fantasy. However, not long after silicon power transistors began to replace germanium types, it became only natural to advantageously exploit the inherent high-frequency capabilities of these transistors; this generally extended into the hundreds of megahertz range. However, the design and implementation of successful amplifiers long remained a black art. Hams, experimenters, and hobbyists could, by patient cut-and-try and individual tailoring, obtain respectable results at ever-higher frequencies. Manufacturers shied away from replacing their tube designs with solid-state amplifiers because they found the transistor RF amplifier to be unpredictable, unstable, unreliable, and not amenable to production requirements.

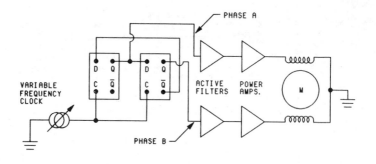

(A) Circuit of digital generator.

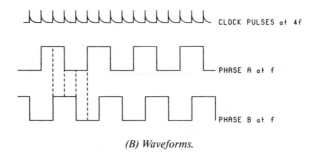

(B) Waveforms.

Fig. 1-17. Digital logic generator for two-phase motor.

It was not until semiconductor firms began to market power transistors especially designed for RF power service that CB equipment, marine, aircraft, and military and space communications systems began to benefit from the obvious attributes of the transistor—small size and weight, easy compliance with vehicle DC power sources, long life, freedom from the need of filament power, and mechanical ruggedness. Thus, it was not enough to produce good power transistors for use in audio work, motor control, regulators, etc., that incidentally also had high-frequency capability. In practice, it was not easy to advantageously utilize such capability. This is because of the difficulty of delivering and extracting power from these devices at the exceedingly low impedance levels involved, which tends to be in the neighborhood of one tenth to several ohms. Further complicating this situation was the effect of lead inductance and package capacitance, neither of which assume importance at low frequencies.

For example, the few nanohenries of inductance involved in the emitter connecting lead produce sufficient degeneration at VHF and UHF frequencies to seriously degrade power gain in common-emitter circuits. And a tiny amount of such lead inductance in the base lead causes regeneration in the common-base circuit and can make the amplifier unstable or even oscillatory. Also, with ordinary low-frequency packaging techniques, the stray capacitance between output and input often makes neutralization mandatory in order to avoid oscillation. Neutralization, however, tends to be a tricky affair because changes in either the driving or loading situation can alter internal transistor parameters enough to provoke oscillation. Also with low-frequency packaging, it proved difficult to achieve broadband operation. All in all, the ability to make silicon transistor chips with intrinsic high-frequency response was a necessary but not sufficient step in the pursuit of solid-state RF power.

Once these matters were realized, the semiconductor manufacturers adopted a wise policy with regard to RF power transistors. They designed packages to optimize RF performance, and they stipulated that such transistors were intended to work from a fifty-ohm driver source to deliver power into a fifty-ohm load. This had very important ramifications. It, of course, stimulated the use of fifty-ohm coaxial cable—good practice in minimizing coupling between output and input circuits. It standardized conditions for specifying the performance of the transistors, reducing the previous tendency for wildly different results from one user to another. And, it removed the black magic from impedance matching, broadbanding, and controlling RF losses. Moreover, the fifty-ohm impedance level worked out to be a good practical compromise for the implementation of strip-line techniques at UHF and microwaves.

An example of one of the things done to make dedicated RF power transistors is depicted in Fig. 1-18. Here, the inductance of the base lead is combined with a small MOS chip capacitor to form an input impedance matching network. This does not necessarily imply that external networks are not necessary. However, the lead inductance now aids in the overall impedance-matching process. Without such a partial network within the package itself, a totally external network could not be very effective in bringing the impedance match to the actual base of the transistor.

Another noteworthy method is the use of special packaging techniques to minimize emitter lead inductance, and to reduce parasitic capacitance between collector and base. Instead of the familiar T03 package commonly used with ordinary power transistors, the stripline opposed emitter package evolved not only to provide a natural interface with UHF stripline techniques, but to accomplish the aforementioned

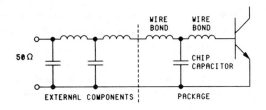

Fig. 1-18. RF power transistors may contain matching networks within their package.

objectives. There are actually two emitter leads and these have much less inductance than the wirebond leads of conventional power transistors. (The concept of inductance in short lengths of conductor may seem far fetched from the vantage point of low-frequency electronics; in the high RF region, it is not a negligible circuit parameter and its practical effect is of the utmost significance.) A family of RF power transistors utilizing the stripline opposed package is shown in Fig. 1-19. A more detailed view of this packaging technique is depicted in Fig. 1-20. Note the materials used for optimum electrical and thermal performance. It can also be seen that the transistor chip contains multiple emitters. This mitigates against a destructive phenomenon when ordinary transistors are operated at RF, known as hotspotting. It can be appreciated that the dedicated RF power transistor is a more sophisticated device than the garden variety type.

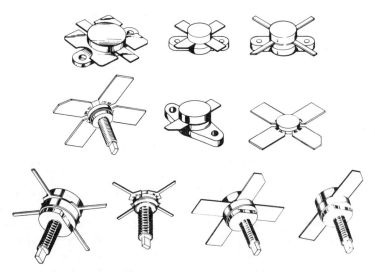

Fig. 1-19. A family of RF power transistors utilizing the stripline opposed emitter package. (Courtesy Motorola Semiconductor Products, Inc.)

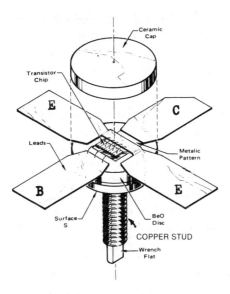

Fig. 1-20. Detailed view of stripline opposed emitter RF power transistor. (Courtesy Motorola Semiconductor Products, Inc.)

The coaxial package shown for the RF power transistor of Fig. 1-21 is another very effective packaging technique. It is particularly useful in the microwave region.

Fig. 1-21. RF power transistor in a coaxial type package. (Courtesy RCA)

RF POWER FROM MOSFETS

The RF bipolar power transistor is now a successful device and is widely used in place of tube amplifiers in the five to several-hundred watt range. Higher power levels are attainable by means of simple power-combining schemes. Despite the realizable performance from these devices, they inherently possess shortcomings which require special attention to avoid malfunctioning or catastrophic destruction. These are as follows:

- Bipolar transistors are subject to thermal runaway, a potentially destructive phenomenon. Once started, the process is regenerative—higher current leads to higher temperature, which in turn leads to higher current, etc. Some units are susceptible to hot spotting, a destructive phenomenon from uneven current distribution in the emitter-base region.
- Bipolar transistors have a restricted safe-operating area of voltage-current combinations because of vulnerability to secondary breakdown. Unlike zener and avalanche breakdown mechanisms, secondary breakdown is destructive.
- Bipolar transistors in RF power applications are subject to low-frequency instability. This is because the low-frequency current gain is generally much greater than the current gain at the operating frequency. Violent low-frequency oscillation can be destructive.
- Bipolar transistors usually cannot be directly paralleled because current is not likely to distribute evenly; any disparity tends to regeneratively worsen, culminating in current hogging by one of the devices. Ballast resistances overcome this, but at the expense of overall efficiency.
- Varactor action in the collector circuit of bipolar transistors increases harmonic generation.
- Bias circuitry can be complex—it is often necessary to thermally track, or otherwise control, the bias voltage.

The RF power MOSFET is virtually free of the above disadvantages. Being a majority carrier device, frequency capability is not limited by minority charge storage effects. The power MOSFET's impedance levels tend to be more "tubelike," making it easier to implement tank circuits and resonant networks at high frequencies and high powers.

Here, again, a marketable product followed on the heels of a dramatic technological breakthrough. It, indeed, remains difficult for some engineers to appreciate that MOS devices are no longer relegated to flea-power applications. The advent of respectable RF power from MOSFETs has forced the makers of bipolar RF devices to accelerated improvement of their devices in order to remain competitive.

There are now complete lines of P-channel power MOSFETs (analogous to PNP transistors) available for complementary-symmetry circuitry in low-frequency power control. Gradually, P-channel RF MOSFETs are also being introduced. (Note, however, that any power MOSFET can be made to work from a reverse polarity DC supply.)

DRAWBACKS IN RF POWER MOSFETS

The natural high-frequency capability of power MOSFETs, together with freedom from the major disadvantages of bipolar transistors appears almost too good to be true. Our intuition stands us in good stead, for the MOSFET *does* have some disadvantages when compared to bipolar transistors for RF service. These can be cited as follows.

The MOSFET is vulnerable to damage or destruction from gate puncture. Overdriving or overbiasing can destroy the thin oxide dielectric of the gate. Even worse, such destruction can be the result of static electricity or leakage currents from the soldering iron. Although protective diodes are incorporated within the package or structure of some power MOSFETs, this technique is not altogether satisfactory for RF MOSFETs.

The saturation voltage (corresponding to $C_{V(sat)}$ for bipolar transistors) tends to be relatively high in RF power MOSFETs. This increases device dissipation and may appreciably degrade efficiency at low DC operating voltage. Continual improvement in this parameter is the order of the day, however.

Special attention must be directed to the relatively high capacitance of the gate input circuit in the power MOSFET. Because of this capacitance, it is usually necessary to drive the MOSFET from a very low impedance source. This is especially true where broadbanding is desirable. Thus the touted high-impedance input of the MOSFET is not a practical reality for RF work. (The input impedance is, however, an order of magnitude higher than in comparable bipolar designs, and it is usually easier to implement practical LC elements.)

Some manufacturers have neither packaged nor specified their otherwise noteworthy power MOSFETs for RF service. For example the giant IRF35X family of power MOSFETs made by International Rectifier Corp. has compelling features for the experimentally inclined. These 150-watt devices operate from 350- to 400-volt DC supplies and therefore display tubelike impedance levels. This leads to easy selection and implementation of resonant circuits. It would appear that these devices may be useful for amateur radio purposes up to five megahertz or so. This, however, is the author's, not the manufacturer's opinion. In general, the shortcoming of RF power MOSFETs is that they have been largely directed towards military and space communications markets, making them more costly than bipolar types for hams, hobbyists, and experimenters.

POWER INTERFACE BETWEEN LOGIC AND ACTION

Microprocessors and other digital logic may be likened to the brains and nerve systems of biological species. There must also be an analogy to the muscles and limbs that exert specified patterns of work on command. Such an analogy may be found in the form of power semiconductors and the motors, solenoids, and other devices they actuate. A widely used power semiconductor for such purposes is the triac. This is a very useful device because powerful AC motors and solenoids can be turned on and off by commands originating in the programmed behavior of the digital circuit. The problem in such an association is the interface between logic and power circuitry.

An obvious approach is the brute-force one utilizing whatever combination of passive and active components is needed to translate logic levels to triac gate signals. A typical circuit arrangement for accomplishing this is shown in Fig. 1-22. Such methods are straightforward and have often been entirely satisfactory. A shortcoming, however, is that there is no electrical isolation between low- and high-power sections of the system. Isolation is desirable, for it helps attain operator safety and it tends to protect the logic circuitry from damaging transients. It would also block a path where transients can back up into the logic gates and cause faulty operation. Power circuits are often the victims of catastrophic failures; a common one is the burn-out of lamp filaments. During such an event, the metallic vapor provides a near short-circuit current path. If this should blow out the triac, the resultant current surges could, in some instances, also damage the logic cir-

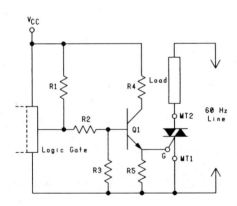

Fig. 1-22. Straightforward way of interfacing digital and power circuits.

cuitry. All things considered, a nonelectrical link between logic and power systems is called for.

Fig. 1-23 depicts a simplified scheme utilizing an optical link between logic and power. Here, an LED actuates a phototriac which, in turn, triggers the main triac. Basically, the phototriac substitutes for the commonly used diac in triac trigger applications. It is obvious that the alluded-to unilateral transmission path is completely realized—no electrical event in the power circuit can be communicated back to the logic system. Although other types of optical isolators can be used, the phototriac is particularly well adapted for triggering the fairly large triacs that are often associated with the load.

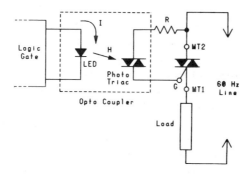

Fig. 1-23. A better way of interfacing digital and power circuits.

BIMOS POWER SWITCH

At certain combinations of voltage and frequency, designers of industrial equipment have long faced a cost and performance dilemma. When it has been advantageous to operate at switching speeds between, say 20 kHz and 200 kHz and control several to several tens of kilowatts in 500- to 1000-volt loads, what is the appropriated power device? The answer directly affects a large market comprising high-frequency bridge inverters for motor drives, uninterruptible power supplies, 440–480 volt off-line equipment, off-line switching power supplies, and welders. Equipment operating from three-phase power lines also tends to fall into this general classification.

Bipolar power transistors capable of good switching performance at

25 kHz or so, are inordinately expensive when one specifies a safe operating area in the neighborhood of 800 volts or higher. A typical bipolar power transistor with 10 ampere capability is found to have a V_{CEO} rating of only 400 volts when used in conventional power switching circuits (in the common-emitter connection). The trade-off for higher V_{CEO} capability is cost. The exotic SOA ratings above 400 or 500 volts are accompanied by exotic prices.

It would be natural to suppose that the power MOSFET is the answer, inasmuch as this newer device outperforms bipolar power transistors in so many applications. The drawback here is that these devices develop wasteful voltage drops when rated for high-voltage service. For example, an 850-volt power MOSFET would drop about 18 volts, and a MOSFET with 1000-volt capability would show a nominal 23-volt drop from drain to source. (These are 10-ampere load situations.) By comparison, the bipolar transistor has a $V_{CE(sat)}$ of only 2.5 volts under the same load conditions. This is a tantalizing situation, for what is needed is a device with the switching and high-voltage capabilities of the power MOSFET, but with the relatively low voltage drop of the bipolar power transistor. (Recall that power dissipation increases as the *square* of the voltage drop across the switching device.)

There are, to be sure, other devices which merit consideration. The GTO (gate turn-off) thyristor can handle large power at a low voltage drop and dissipation. This device shows considerable promise, but best applications are in low-frequency circuits. Also, many designers do not like to contend with the commutation problems that sometimes accompany all thyristor circuits. That is also why inverter-rated SCRs tend to take second place to bipolar transistors, Darlingtons, and MOSFETs whenever these devices can be pressed into service as power switches.

Fortunately, a power switch is now available for reliable and cost-effective use in the voltage and frequency combinations where other devices tend to produce awkward compromises. It is a composite circuit called a BIMOS. The basic arrangement is shown in Fig. 1-24.

In this arrangement, the power switch exhibits the high-voltage and high-frequency capabilities of the bipolar transistor together with the ease of drive of the power MOSFET. Because relatively inexpensive devices are used, the overall cost is considerably less than single devices of either type with equivalent speed and voltage ratings. Note: BIMOS is a term used by International Rectifier Corp. This company has done original development work in combining the practical features of power MOSFETs and bipolar power transistors. Their line of power MOSFETs is trademarked HEXFETs, but in this book, we will use the generic term—MOSFET.

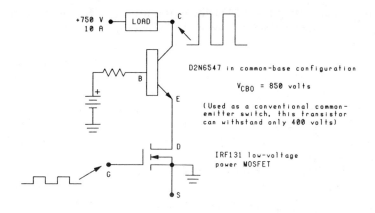

Fig. 1-24. The BIMOS is a composite circuit power switch with compelling features. (Courtesy International Rectifier Corp.)

In the simple circuit of Fig. 1-24, the 10-ampere voltage drop across the bipolar transistor is 2.5 volts; with the same load, the power MOSFET develops a drop of 4.3 volts. Accordingly, the total voltage drop of the composite circuit is 6.8 volts. So power dissipation is low and is shared by two devices. Frequency capability is about what one would get from just a common-base bipolar circuit, but ease of drive is very nearly that of a MOSFET by itself. The circuit of the BIMOS power switch is known as a cascode configuration and has seen various services in both tube and transistor applications. The fortuitous performance and cost factors displayed by the power-switch application is a relatively recent discovery. Monolithic integration of the two devices unfortunately poses some manufacturing problems. A single-package hybrid version can be expected to become available, however.

THREE-TERMINAL VOLTAGE REGULATOR

The three-terminal voltage regulator has deservedly enjoyed widespread popularity in all manners of electronics applications. These devices have greatly simplified the implementation of voltage regulation, and have at the same time dramatically reduced parts count and overall cost. And by making practical the concept of regulation close to the circuitry where needed, improved performance has been realized in many different equipments. The AC-bypassing action of these regulators has greatly alleviated problems with interaction between stages and circuit sections because of the mutual impedance of unregulated DC sources, and of connecting cables. For the most part, these IC regulators have carried current ratings of between one and eight amperes, and have been available in output voltage ratings of between five and thirty-five volts. Both positive and negative types have been marketed. Some, instead of having fixed output voltages, are designed to be adjustable.

The three-terminal voltage regulator was the first widely produced power IC. As such, it paved the way for other useful ICs combining "brains and brawn" on a single monolithically integrated chip. Although the three-terminal regulator was initially intended to be used for its specifically designated circuit function, its inherent versatility became manifest early in the scheme of things. Thus, it became common for manufacturers to show how these devices could be used in other than a narrowly specified way. For example, they are readily made to operate as current regulators; as such, they can provide either fixed or variable currents to a load. By associating them with an external power transistor, their current, voltage, or power capability can be extended. By appropriate circuit techniques, a positive regulator can be used to regulate a negative output voltage, and vice versa. Even though best results are achieved in variable-voltage regulation by using a regulator designated for such service, a fixed-voltage type can be made to perform acceptably well for many applications. Finally, these linear regulators work exceedingly well as switching regulators in relatively simple circuits.

Despite their apparent simplicity, many three-terminal regulators contain up to several tens of active devices and provide operational features previously found only on deluxe laboratory supplies. These include thermal-overload shutdown, SOA protection of the series-pass element, current limiting, and stable temperature-compensated voltage references. Regulation specifications are often tighter than needed; line regulation is frequently found to be in the .005–.01% per volt range and load regulation is often specified in the .05–0.1% per volt range. Ripple rejection, which indicates the device's ability to act as an electronic filter, is usually in the 60–80 db range—higher than is easily attainable in series-pass regulators designed around discrete devices.

Because the three-terminal voltage regulator is a self-contained subsystem, it might be supposed that it is only necessary to connect it between the unregulated DC source and the load, and assume it need be given no further thought. This is almost true most of the time; the maker has done his utmost to make these devices "idiot-proof" and quickly applicable to a wide variety of situations. However, success in the practical implementation of the three-terminal regulator is best achieved by awareness of its vulnerabilities and by the observance of some basic ground rules.

The simple three-terminal regulator circuit is shown in Fig. 1-25. An even simpler implementation sometimes results from eliminating capacitors C1 and C2. In general, however, these capacitors are either essential or desirable. Input capacitor C2 is often prescribed as a ceramic or solid tantalum type in the capacity range of 0.1 μF to several microfarads. This capacitor must be physically situated as close as possible to the regulator. In those instances where it is permissible to dispense with C2, it is because the connecting lead from the filter capacitor of the unregulated DC source is short. The prime function of C2 is to make the regulator stable, and particularly to prevent high-frequency oscillation. For this reason, aluminum electrolytic capacitors such as are commonly used for power-supply filtering are not always trustworthy. As a rule of thumb, the aluminum electrolytic capacitor would have to be twenty-five times or more the size of a ceramic or solid tantalum capacitor for equivalent high-frequency bypassing effect. Even the low-frequency impedance of the aluminum filter capacitor can be inadequate for satisfactory bypassing; this is because of the ESR (effective series resistance) of these capacitors.

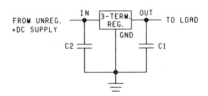

Fig. 1-25. Generalized implementation of the three-terminal voltage regulator.

Although ceramic capacitors are universally used for bypassing higher frequencies, some types seem to suffer an appreciable decrease in capacitance around a half megahertz, or so. Although this behavior will rarely cause trouble, it occasionally is the source of mysterious instability. All things considered, the solid tantalum capacitor is a good choice for C2. Where size considerations do not pose problems, mylar capacitors are also excellent.

Output capacitor C1, also helps stabilize the regulator against oscillation or instability. Generally, however, its prime function is to improve the transient response. It, too, can be dispensed with under some conditions. If, for example, the load has such a capacitor, and is located close to the regulator, C1 might not be needed. If the regulator exhibits stability and transient response is not an important factor, C1 would not be needed. Some regulators tend towards instability when C1, or its equivalent, is relatively small—say in the 500 pF to 5000 pF range. Thus, C1 is commonly stipulated to be in the 0.25 μF to several microfarads range. Here, again, ceramic and solid tantalum capacitors, as well as mylar types generally serve well.

THREE-TERMINAL ADJUSTABLE VOLTAGE REGULATOR

The basic circuit for obtaining adjustable output voltage from a three-terminal voltage regulator is shown in Fig. 1-26. This circuit configuration will work satisfactorily in some applications with many of the fixed-voltage types, but best performance will be achieved with specially designated adjustable regulators, such as the LM117, LM137, LM138, and LM150 family of regulators. Typical values for R1 lie in the 100 to 300 ohm range; R2 is commonly a 5K potentiometer connected in rheostat fashion. For increased ripple rejection, the optional capacitor, C3, may be added; 10-μF aluminum electrolytic types generally serve this purpose, but heed should be paid to polarity. If this capacitor is used, the regulator becomes vulnerable to damage from certain fault conditions—this will be discussed subsequently. This circuit will not allow adjustment all the way down to zero voltage. However, a minimum output of 1.2 volts is usually attainable, and is low enough for most practical use of an adjustable-voltage source.

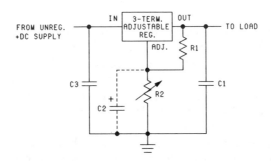

Fig. 1-26. Generalized implementation of the three-terminal adjustable regulator.

In order to be able to adjust to zero voltage, an arrangement such as depicted in Fig. 1-27 must be employed. Here, the ADJ terminal of the regulator connects to a negative 1.2-volt source instead of ground. The LM113 may be considered, for practical purposes, a synthesized 1.2-volt zener diode. An additional requirement is an auxiliary low-current negative 10-volt source. As shown, this modification allows a 0–30 volt adjustment range of the regulated output voltage.

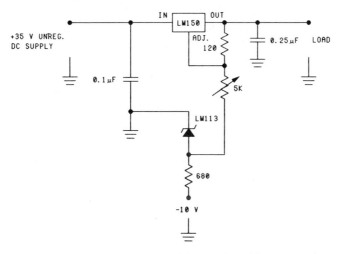

Fig. 1-27. Example of technique for obtaining adjustable output voltage down to zero.

USING DIODES TO PROTECT THREE-TERMINAL REGULATORS

Regulator ICs are vulnerable to damage or destruction from certain fault conditions. Consider the circuit of Fig. 1-26. If the input is shorted to ground, capacitor C1 will discharge into the output terminal of the regulator. Whether this portends danger to the regulator depends on the way the internal circuitry of the regulator is fabricated, upon the size of C1, and upon the voltage across C1. Generally, a high-current regulator which delivers low output voltage, together with a small output capacitor represents a combination of conditions in which the regulator is not endangered by such a fault. A combination of opposite conditions can, however, endanger the regulator.

Somewhat similarly, if either the input or output is shorted to ground, ADJ bypass capacitor C2 discharges into the ADJ terminal and can cause catastrophic destruction. Here again, much depends upon the size of the capacitor, the nature of the IC, and the voltage across the capacitor. The 10-μF capacitor often used in this circuit is not large enough to endanger many IC regulators, especially if five volts or so is the output voltage. Sometimes, for the sake of increased ripple rejection, capacitor C2 is made much larger than 10 μF, whereupon its sudden discharge current can pose a threat to the regulator.

It is possible to protect against the possible consequences of such circuit faults by adding diodes, as shown in Fig 1-28. Here, diode D1 protects against discharge from capacitor C1; diode D2 protects against discharge from capacitor C2. These diodes can be ordinary high-current types such as the 1N4002. Note that the diodes are so polarized as to exert no effect on the normal operation of the regulator.

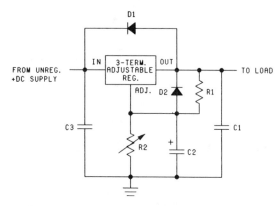

Fig. 1-28. A three-terminal regulator circuit with protective diodes.

EASY WAY TO ZAP THE IC VOLTAGE REGULATOR

There remains yet another way in which three-terminal and other IC regulators can be destroyed, as the saying goes "without half-trying"; that is, by interrupting the ground or ADJ connection. When this is done, the output voltage rises to approximately the input voltage from the unregulated DC supply. This, in many instances, damages or destroys devices or components in the load circuit. And, when the open connection is again completed, the IC regulator itself can suffer destruction. An obvious reaction to this information is, of course, that there is no reason to expect such a fault condition—that such an occurrence has a very remote probability of taking place.

Superficially, this reasoning appears sound. However, it neglects consideration of a very common test technique during breadboard evaluation, and also during troubleshooting procedures. This entails either the removal and plugging in of the regulator IC, or of the card on which it is mounted, while power is on. It doesn't require much imagination to visualize that all three terminals of the regulator, or the corresponding pins of the card connector probably do not make simultaneous contact during the plugging-in act. Accordingly, one can expect that sometimes the ground or ADJ connection will be the last in the contacting sequence. This being the case, the precaution against mysterious trouble of this nature is simple enough: *Do not remove or replace either the regulator IC or the card on which it is mounted while power is on!*

THERMAL HINTS AND KINKS

This treatise does not detail the subject of heat removal on the premise that the reader has both the awareness and the practical knowledge to implement successful hardware for keeping junction temperatures of power semiconductors within safe limits. Engineers accomplish this via mathematical treatment of the thermal circuit. Hobbyists generally fly by the seat of their pants—by intuition, cut and try, and comparison with known or previous situations, they generally attain practical success. Fortunately, there is little penalty for providing more than enough heat removal other than usage of a little more space and a few cents of extra expenditure for material. A brief discussion of thermal matters as they pertain to IC regulators is relevant, however. This is because most of these regulators have internal thermal-shutdown or current-limiting provisions. Thus, it might be construed that they are immune to damage from overload.

The thermal protection inherent in these IC regulators does, indeed, usually prevent catastrophic destruction from overload. It is not wise, however, to operate them for long periods, or on a repetitive basis at their temperature limit. Moreover, it is good practice from a reliability standpoint to ensure that normal operation occurs well below the shutdown or limiting level of current. To accomplish this, attention must be given to heat removal. It is often desirable to utilize the copper foil of the PC board as a heat sink. Whether this is feasible will depend upon individual conditions. A five-ampere regulator can generally be expected to pose a more difficult thermal problem than a one-ampere regulator, and is more likely to require an external heat sink. In all cases, however, the problem of heat removal can be eased by not using a higher input voltage than is needed under worst operating conditions. Most IC regulators require about two volts differential in order to operate properly. Keeping this in mind, one has to appraise the situation when maximum load demand and minimum line voltage, i.e., unregulated DC input voltage, exist simultaneously.

One should strive for optimum efficiency of the three methods of heat removal. These are conduction, convection, and radiation. Conduction is governed by nature of the material and its volumetric mass. Thus, copper is better than aluminum, and a thick copper sheet on a PC board is more effective than a thin one. Convection is greatly influenced by the availability of circulating air. This has to do with physical placement, obstructions, and access to cooler ambient air. Radiation is a function of surface conditions. For example, a smooth shiny surface—even of copper—will not radiate heat well. Conversely, rough blackened surfaces perform efficiently as heat radiators. The emissivity of various surfaces is depicted in Table 1-1. The higher the emissivity, the more efficient is radiation. (Do not be confused by the polished reflectors used in electric heaters. The reflector does not radiate heat from its own interior; rather, it merely reverses the direction of heat from the heating element.) Representative heat sinks are listed in Table 1-2. These are particularly applicable to three-terminal and other IC regulators. For best convective effect, fins should be vertical. Silicone grease should be applied to the junction between the IC and heat sink. Here, the lower the thermal resistance θ_{SA}, the more effective is the transference of heat to the surrounding air.

Table 1-1. The Radiation Efficiency of Various Surfaces
(Courtesy National Semiconductor Corp.)

SURFACE	EMISSIVITY, E
Polished Aluminum	0.05
Polished Copper	0.07
Rolled Sheet Steel	0.66
Oxidized Copper	0.70
Black Anodized Aluminum	0.7 –0.9
Black Air Drying Enamel	0.85–0.91
Dark Varnish	0.89–0.93
Black Oil Paint	0.92–0.96

WIRING HINTS FOR THREE-TERMINAL REGULATORS

Because the current being processed is DC, it is commonly held that it makes little difference how the three-terminal regulator is wired or connected from a geometrical standpoint. It is, according to this notion, much like a doorbell circuit—about the same operation will ensue no matter what goes where as long as the basic current path is correct. This is far from the truth with regard to three-terminal regulators. Because of the high currents, high amplification, and close regulation expected, the tiny resistance of short lengths of connecting lead can make itself felt in performance. Indeed, more often than not, the true capability of these ICs is not realized because no effort is made to optimize the wiring pattern. Even though a draftsman and a technician may accept the connection diagram of a regulator, the measurements later made of actual performance may mysteriously fall well below both specifications and expectations.

In the semipictorial diagram shown in Fig. 1-29A, two very important construction techniques are depicted. First, a single ground point is used. This ground point should be as close as possible to the load. This technique avoids ground loops which easily couple from one section of the regulator circuit to another, and often cause excessive output ripple, instability, and degradation of regulation. Second, R1, the "set" resistor, should be connected as close to the output terminal as possible. If this connection, instead, is made near the load, the effect of lead resistance will be multiplied and load regulation can be seriously degraded.

In Fig. 1-29B the lead resistance of the connection between GND and ground is often ignored and the filter capacitor of the unregulated DC supply is connected directly to the GND terminal of the regulator. "Doorbell-circuit" psychology would suggest that this is the same as making this connection to ground, as indicated by the dashed line. This is not so, however; if the connection is made via the wrong method, it is possible to induce considerable ripple into the load circuit. An attempt to remedy this type of malperformance by increasing the size of the filter capacitor tends to actually aggravate the ripple in the output of the regulator.

If, because of practical reasons, the ideal connection patterns cannot be attained, they should be approached as closely as is feasible. It is always helpful, too, to make the critical lead resistances very low; this can be done by means of large conductors or multiple conductors. For example, some regulators use the TO-3 package, with the case being the GND connection. With such regulators, it is convenient to provide *two*

Table 1-2. Representative Heat Sinks (Courtesy National Semiconductor Corp.)

Θ_{SA} Approx (°C/W)	Manufacturer & Type	Θ_{SA} Approx (°C/W)	Manufacturer & Type	Θ_{SA} Approx (°C/W)	Manufacturer & Type
For TO-202 Packages		For TO-5 Packages		For TO-3 Packages	
12.5 – 14.2	Staver V4-3-192	12	Thermalloy 1101, 1103 Series	0.4 (9" length)	Thermalloy (Extruded) 6590 Series
13	Staver V5-1	12 – 16	Wakefield 260-5 Series		
15.1 – 17.2	Staver V4-3-128	15	Staver V3A-5	0.4 – 0.5 (6" length)	Thermalloy (Extruded) 6660, 6560 Series
19	Thermalloy 6106 Series	22	Thermalloy 1116, 1121, 1123 Series	0.56 – 3.0	Wakefield 400 Series
20	Staver V6-2	22	Thermalloy 1130, 1131, 1132 Series	0.6 (7.5" length)	Thermalloy (Extruded) 6470 Series
25	Thermalloy 6107 Series				
37	IERC PA1-7CB with PVC-1B Clip				
40 – 42	Staver F7-3	24	Staver F5-5C	0.7 – 1.2 (5 – 5.5" length)	Thermalloy (Extruded) 6423, 6443, 6441, 6450 Series
40 – 43	Staver F7-2	26 – 30	IERC Thermal Links		
42	IERC PA2-7CB with PVC-1B Clip	27 – 83	Wakefield 200 Series	1.0 – 5.4 (3" length)	Thermalloy (Extruded) 6427, 6500, 6123, 6401, 6403, 6421, 6463, 6176, 6129, 6141, 6169, 6135, 6442 Series
42 – 44	Staver F7-1	28	Staver F5-5B		
		30	Thermalloy 2227 Series		
For TO-220 Packages		34	Thermalloy 2228 Series		
4.2	IERC HP3 Series	35	IERC Clip Mount Thermal Link	1.9	IERC E2 Series (Extruded)
5 – 6	IERC HP1 Series	39	Thermalloy 2215 Series	2.1	IERC E1, E3 Series (Extruded)

Table 1-2 (Cont.). Representative Heat Sinks (Courtesy National Semiconductor Corp.)

Θ_{SA} Approx (°C/W)	Manufacturer & Type	Θ_{SA} Approx (°C/W)	Manufacturer & Type	Θ_{SA} Approx (°C/W)	Manufacturer & Type
6.4	Staver V3-7-225	42	Staver F5-5A	2.3 – 4.7	Wakefield 600 Series
6.5 – 7.5	IERC VP Series	45 – 65	Wakefield 296 Series	4.2	IERC HP3 Series
8.1	Staver V3-5	46	Staver F6-5, F6-5L	4.5	Staver V3-5-2
8.8	Staver V3-7-96	50	Thermalloy 2225 Series	5 – 6	IERC HP3 Series
9.5	Staver V3-3	50 – 55	IERC Fan Tops	5.2 – 6.2	Thermalloy 6103 Series
10	Thermalloy 6032, 6034 Series	51	Thermalloy 2205 Series	5.6	Staver V3-3-2
12.5 – 14.2	Staver V4-3-192	53	Thermalloy 2211 Series	5.8 – 7.9	Thermalloy 6001 Series
13	Staver V5-1	55	Thermalloy 2210 Series	5.9 – 10	Wakefield 680 Series
15	Thermalloy 6030 Series	56	Thermalloy 1129 Series	6	Wakefield 390 Series
15.1 – 17.2	Staver V4-3-128	58	Thermalloy 2230, 2235 Series	6.4	Staver V3-7-224
16	Thermalloy 6106 Series	60	Thermalloy 2226 Series	6.5 – 7.5	IERC UP Series
18	Thermalloy 6107 Series	68	Staver F1-5	8	Staver V1-5
19	IERC PB Series	72	Thermalloy 1115 Series	8.1	Staver V3-5
20	Staver V6-2			8.8	Staver V3-7-96
25	IERC PA Series			8.8 – 14.4	Thermalloy 6013 Series
26	Thermalloy 6025 Series			9.5	Staver V3-3
For TO-92 Packages				9.5 – 10.5	IERC LA Series
30	Staver F2-7			9.8 – 13.9	Wakefield 630 Series
46	Staver F5-7A, F5-8-1			10	Staver V1-3
50	IERC RUR Series			13	Thermalloy 6117
57	Staver F5-7D				
65	IERC RU Series				
72	Staver F1-7				
85	Thermalloy 2224 Series				

Staver Co, Inc: 41-51 N. Saxon Ave, Bay Shore, NY 11706
IERC: 135 W. Magnolia Blvd, Burbank, CA 91502
Thermalloy: PO Box 34829, 2021 W. Valley View Ln, Dallas TX
Wakefield Engin Ind: Wakefield MA 01880

paths from case to circuit ground by attaching lugs to both mounting screws.

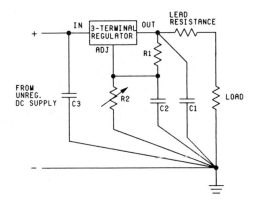

(A) Right method. Single-point connection for ground paths avoids malperformance from mutual ground loops.

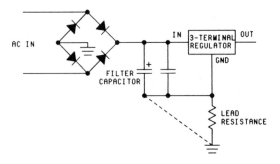

(B) Wrong method. Filter capacitor in unregulated supply should connect to the ground point (dashed line), not to the GND terminal of the regulator.

Fig. 1-29. Wiring methods can affect performance of three-terminal regulators.

CHAPTER 2

Solid-State Amplifiers

Power amplifiers configured around discrete devices and fairly ordinary silicon output transistors have yielded surprisingly good performance; yet, there has been room for improvements not readily achievable with the traditional devices. Consider, for example, the technique of negative feedback, one of the important circuitry "weapons" for extending frequency response and ironing out the effects of nonlinearities. The mere accomplishment of reasonably flat frequency response, say from 20 Hz to 20 kHz, never told the whole story with regard to high-fidelity reproduction.

Missing from the response specifications was what was happening beyond 20 kHz. At first glance, this seems inconsequential because most people's hearing tends to go into high attenuation after 17 kHz or so. However, when an audio amplifier exhibits nonlinearity or positive feedback at higher frequencies, the amplifier's quick transient response is degraded; in addition, beat frequencies and intermodulation distortion are sure to creep into the audio band itself. As a rule of thumb, it would be wise to have audio amplifiers be flat out to about 200 kHz. But there is more to this requirement than meets the eye. Stereo equipment manufacturers have long assumed that the shortcomings of the nonfeedback amplifier could be overcome with sufficient feedback. The premise has been that if a little feedback is good, more must naturally be better.

However, engineers in more exacting fields such as telephony or instrumentation have always been aware that it is best to start originally from the best nonfeedback amplifier achievable before piling on the negative feedback. We shouldn't be too severe with the stereo manufacturers, however; they really have done what was possible at a reasonable cost. The power transistors that have been used in stereos have been far from ideal devices with regard to linearity and frequency response. Then, there has been the matter of crossover distortion from the class AB push-pull output stages. As if this wasn't enough, further departure from linearity came from the commonly used quasi-complementary-symmetry circuits.

These and other shortcomings can be largely overcome by using some of the newer power devices. And almost everything that has been said with regard to stereo amplifiers applies also to servo amplifiers, especially where high accuracy and low overshoot are performance goals. Oscillation, hunting, and other mechanical aberrations plague servo systems with too much patchwork negative feedback. Here again, some of the modern power devices tend to produce better results than many of the traditional brute-force power transistors.

SIMPLE INTERCOMS WITH AUDIO POWER ICS

Power op amps, especially those design dedicated for use in audio applications, are a natural for simple but effective intercommunications systems. Such intercoms are essential in industrial and commercial environments, but find considerable residential use also. Further simplifying practical implementation is the fact that the common dynamic speaker also functions exceedingly well as a microphone. The easiest intercom system to build, and often to install, is the master-slave format. In this scheme, the operator at the master station actuates a mode switch to determine when he chooses to talk or listen. The person at the slave, or remote station then talks when given the opportunity to do so; i.e., when the mode switch at the master station is placed in its listen position.

Fig. 2-1 shows two intercom systems utilizing audio ICs. All circuitry, including the master speaker/microphone, is located at the master station; the slave speaker/microphone is located at the slave or remote station and is not encumbered with any additional circuitry. In Fig. 2-1A, the LM388 IC can deliver in excess of 1.5 watts to an 8-ohm speaker. A 10- to 12-volt power supply is suitable. Class AB operation is used by the output stage of this IC so that the quiescent current is quite low, about 18 milliamps.

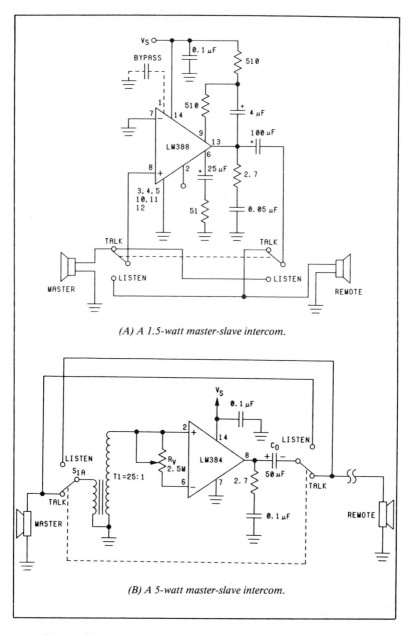

(A) A 1.5-watt master-slave intercom.

(B) A 5-watt master-slave intercom.

Fig. 2-1. Simple intercom systems using power ICs. (Courtesy National Semiconductor Corp.)

The 5-watt intercom shown in Fig. 2-1B is actually more typical of such single IC circuits. It is similar to that of Fig. 2-1A, but a step-up transformer is inserted in the input circuit when either station employs its speaker as a microphone in the talk mode. This is needed because of the lower gain of the more powerful ICs. A 5- or 6-volt filament transformer will probably serve this purpose if operation is limited to voice transmission. The LM384 IC will operate well from a nominal 25-volt DC supply. Approximately four or five square inches of copper foil on a PC board will function as a satisfactory heat sink.

Both of the intercoms just discussed will work with 4-, 8-, or 16-ohm speakers. More available power but lower power gain will be forthcoming from the lower-impedance speakers. However, higher-impedance speakers serve as more sensitive microphones. Eight-ohm speakers should prove to be a good compromise. If 4-ohm speakers are used with the LM384, a clip-on heat sink may be necessary. A relatively large-diameter speaker best serves the acoustical situation likely to prevail at the remote station; at the master station, the operator is close for both talking and listening, so a small speaker should prove suitable.

OPTOISOLATORS AS INTERSTAGE COUPLING ELEMENTS

Optoisolators are very useful devices for many circuit applications. They are often found in feedback loops of switching power supplies, and have found extensive application in the transmission and processing of digital data. They also serve well where it is desirable to isolate control-panel switches and control knobs from the lethal high-voltage systems being controlled. In these applications, they are often found to be less costly and less demanding of space than are transformers. Optoisolators commonly employ an LED infrared emitter and a variety of photosensitive detector elements, including photodiodes, phototransistors, photo-Darlingtons, photo-FETs, and photothyristors. High-voltage isolation capability stems from the physical separation between emitter and detector. Notwithstanding their widespread uses, relatively little has been done with these devices for application as a linear transducer or coupling agent. At first thought it would appear that the optocoupler would make an ideal interstage coupler in audio amplifier systems. This would resolve conflicts in bias and DC voltage levels, and would be expected to exhibit flatter frequency response than either transformers or capacitors.

Qualitatively, this line of reasoning points in the right direction. However, practical implementation has been held in check by problems

in li … dependence, and a general transfer function
that … lesigners as being sufficiently predictable. This
has b … trating, inasmuch as the optoisolator's ability
to in … s can sometimes be very useful. For example,
in int … metimes happens that considerable AC hum is
picke … way industrial and commercial utility installations … thin a building. It fortunately turns out that if
passa … not required, the optoisolator can be made to
provi … ty for some audio-coupling applications, such
as int … rocessing sections of amateur-radio transmitters. T … nplementations is to operate within a relatively
linear … solator's transfer characteristics, and to stabilize th … D current.

Th … Fig. 2-2 provides good linearity and a reasonable c … . The 10-milliampere constant-current source
need r … two-terminal device. For example, the single-
transis … tor depicted in Fig. 1-3C is suitable. Many
three-t … gulator ICs can be used as constant-current
source … be necessary in all applications to use separate
5-volt … parate ground systems.

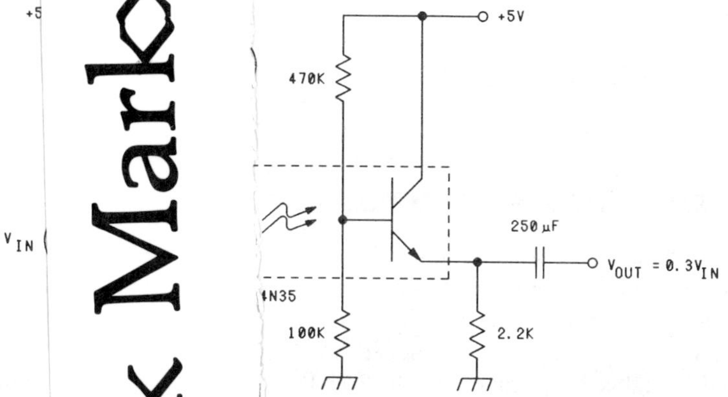

Fig. … ing a near linear transfer of audio signal with … r. (Courtesy General Electric Company)

An a … tor scheme for the interstage coupling of audio sign … 2-3. Here, current stabilization in the LED
circuit i … ferential amplifier. Although no DC blocking capa … heir use will be called for in many applications. O … it operates in much the same manner as the

circuit of Fig. 2-2. The unmodulated LED current is governed by the resistor labeled bias. Empirical determination of this resistor is necessary for the attainment of optimum linearity.

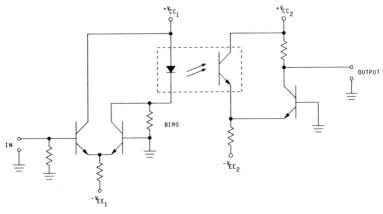

Fig. 2-3. Alternative scheme for near linear coupling of audio signals through an optoisolator.

When either of these circuits is used without the DC blocking capacitors, the low frequency response extends all the way down to zero (DC). However, it is much more difficult to obtain proportional transfer of the DC component than the AC portion of the signal, and temperature effects then tend to make operation somewhat erratic.

GETTING THE BEST FROM MOSFET POWER

The high impedance input characteristic of the power MOSFET has excited a great deal of enthusiasm with designers, for this device also possesses high frequency capability. At DC and audio frequencies, easy drive and flat response are, indeed, readily realized. One must realize that at higher frequencies this device has a high input capacitance, and the effect of this can be exacerbated by the Miller effect. Thus, at higher frequencies, the power MOSFET no longer appears to be a near infinite input impedance device; moreover, its frequency response will roll off long before the actual amplifying capability of the device has suffered. The early frequency roll off actually takes place in the driver output circuit.

What is needed, then, is a very low impedance driver. With such a driver, the input capacitance of the MOSFET will have minimal deleterious effect and wideband performance can be attained. If one can devise a low impedance drive circuit and still retain the high impedance input characteristic of MOSFET devices, a very useful amplifying system will result. This is particularly true if such performance can be coupled with the high output capability of the power MOSFET. These statements may initially appear to involve contradictions, but the circuit shown in Fig. 2-4 readily provides all of the desirable operating parameters, i.e., high-impedance input, wideband frequency response, and high-power output.

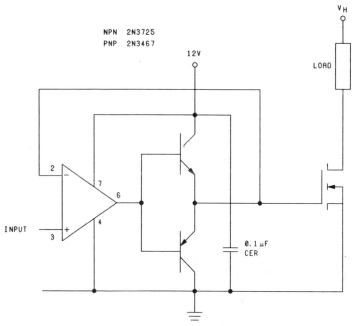

Fig. 2-4. Driver circuit allows both a high-frequency response and high-impedance input from a MOSFET amplifier. (Courtesy International Rectifier Corp.)

In Fig. 2-4, the input device is an operational amplifier with a MOSFET input stage. The input capacitance is approximately 4 pF, compared to 700 pF or so for a power MOSFET. The intermediate stage is a complementary-symmetry, emitter-follower circuit for further reducing the driving impedance seen by the power MOSFET stage. Because of the feedback connection, the first two stages perform as a voltage follower. This feedback technique makes the output impedance of the complementary-symmetry stage lower than it would otherwise be. So, in effect, the power MOSFET output stage is driven by an exceedingly low impedance source, and can thereby deliver wideband frequency response. Depending somewhat on the selection of the bipolar driving transistors and the power MOSFET, an eight to twelve megahertz passband can be realized with this scheme. Of course, the load must have suitable wideband response in order to achieve an overall flat response from the amplifying system.

A TV SOUND CHANNEL IC

Workers in the power-control field have become accustomed to the trend to accommodate both high- and low-power devices on a single chip. The Darlington power amplifier was the first example of monolithic integration of driver and output transistor in an IC module. Transistor arrays and more complex integrations followed, often involving several stages, together with auxiliary stages, such as voltage regulators, protective circuits, etc. A classic example was the three-terminal voltage regulator. As sophisticated as some of these ICs were, they generally did not process widely different kinds of signals, such as those encountered in radio and TV circuits. It was long felt that a diverse mix of functions on a single chip would lead to compromised performance, and a tendency towards various types of instabilities from inadvertent cross-talk and unintended feedback paths.

This is no longer the case, as may be witnessed by inspection of late-model TV sets. In place of the numerous discrete stages and high parts count, one now sees a relatively small number of multiple-function ICs and a dramatic reduction in associated parts. An example of such an IC is the Sprague ULN-2290B TV sound channel. In addition to an audio-output stage with 4-watt capability, this IC contains the audio preamplifier, a regulated power supply, FM detector, and a six-stage limiting IF amplifier. It also contains or has pinout provisions for important circuit functions, such as low-pass filtering following IF amplification, deemphasis, IF amplifier bias, audio feedback, and volume control. The input impedance of the IF amplifier is sufficiently high to enable optimum use of either a ceramic filter or a tuned circuit. The functional block diagram of this IC is shown in Fig. 2-5.

The way this IC is used in a TV receiver is shown in Fig. 2-6. Note that this device, in association with a relatively small number of external components, dramatically simplifies the circuit layout of a TV set. When used in conjunction with other dedicated ICs, the overall savings

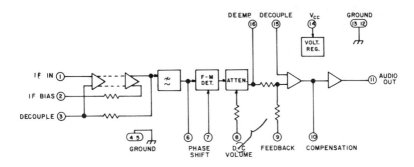

Fig. 2-5. Functional block diagram of the Sprague ULN-2290B TV sound channel IC. (Courtesy Sprague Electric Company)

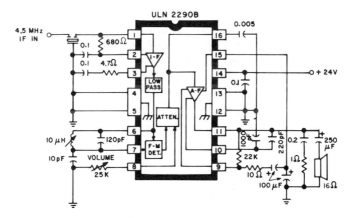

Fig. 2-6. TV implementation of Sprague ULN-2290B TV sound channel IC. (Courtesy Sprague Electric Company)

in assembly labor provides the consumer with much more performance per dollar than was hitherto attainable. Note that connection terminals 4 and 5, as well as 12 and 13, are at ground potential and are in the form of tabs. These tabs can be conveniently provided with simple heat sinks.

ICs of this type have already given a good account of themselves with regard to performance and reliability. However, considerable trouble has been experienced with the sockets. It behooves the designer to carefully consider the mechanical integrity of sockets, and to focus attention on contact materials with good resistance to corrosion; otherwise, the multifunctional IC is not given a fair chance to perform within its capabilities.

DEDICATED IC LOW-COST AUDIO SYSTEMS

Millions of small AM radios, phonographs, intercoms, and other audio equipment have been manufactured with output power levels in the several-watt range. For the large but nondemanding sector of the consumer market, such output power has proven to be adequate. Among the multitudes of satisfied consumers have been TV viewers as well. Often an otherwise elaborate and sophisticated TV chassis will be equipped with a tiny speaker driven by an audio output stage with a couple of watts capability. Of course, the prime explanation of this hinges on cost and competition. But, the other side of the story is that minimal power audio can be satisfactory, particularly if the quality of sound reproduction is good.

The Sprague ULN-2280B is a monolithically integrated 2.5-watt audio system requiring a minimal number of external passive components. This IC contains 23 transistors. The savings in expense, construction effort, and debugging time are compelling reasons for choosing such an IC in place of the extensive parts count of discrete devices and supporting circuitry.

Fig. 2-7 depicts three applications of this audio IC. To assure satisfaction with their products, manufacturers have exerted efforts beyond the mere placing of such sophisticated devices on the market; additionally, they have developed peripheral circuitry to bring out the best possible results from their ICs. This relieves the user of unnecessary design drudgery, as well as experimentation that could result in accidental

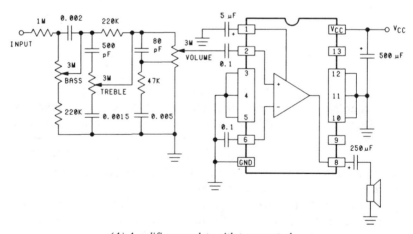

(A) Amplifier complete with tone controls.

Fig. 2-7. Three easily implemented audio applications

damage. The three audio applications shown all use input tone controls. Base boost is a useful expedient for audio systems with small speakers, inasmuch as partial compensation for the small speaker's early low-frequency rolloff is thereby attained. (The hi-fi enthusiast will surely object to such an approach, but such elite hobbyists would hardly be expected to be interested in a 2.5-watt audio amplifier in the first place.)

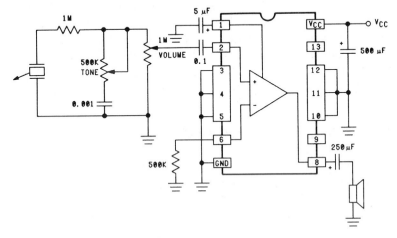

(B) Low-cost phonograph.

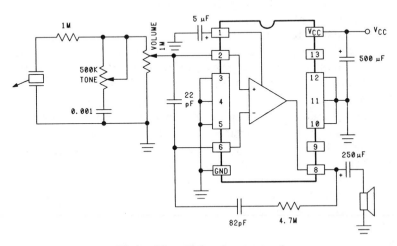

(C) Amplifier with base-boost network.

of the ULN 228 IC. (Courtesy Sprague Electric Company)

The ULN-2280B can deliver 2.5 watts into either an 8- or a 16-ohm speaker. For 8-ohm speaker operation, the prescribed power supply voltage is about 17 volts. For 16-ohm speakers, the power supply should develop 22 volts. In both instances, total harmonic distortion will be close to 3%. Somewhat more power can be extracted if 10% of THD is permissible (as it might be for voice only).

The ULN-2280B has a high-impedance input and a fixed voltage gain of 34 dB. An inexpensive heat sink suffices, the recommended one being a Staver V-8, which clips neatly onto the IC package. Thermal overload protection is incorporated in the internal circuit so that output current is temperature limited. The IC is also self-protected against short circuiting of the AC output. All things considered, dedicated ICs of this type are excellent devices for the economy-minded hobbyist bent on achieving rapid and predictable results.

TWO STEREO CHANNELS ON A CHIP

Another low-power audio amplifier is of interest because it closely approaches the ideal system concept—the system on a chip alluded to in Chapter 1. The stereo-amplifier circuit shown in Fig. 2-8 actually comprises a single IC, the National Semiconductor LM379. This IC is, in essence, a dual-power op amp. It is representative of the modern trend to combine signal processing and high-power capability in a single module. Whereas 6 watts per channel wouldn't impress certain audio enthusiasts as high power, those acquainted with several hundred milliwatt monolithic ICs will accord the LM379 IC a healthy respect. This power level is very satisfactory for stereo phonographs, AM-FM radio receivers, tape recorders and players, and intercoms. It is also quite possible to upgrade the sound system in many TV sets with this simple audio amplifier.

As seen in Fig. 2-8, it is only necessary to associate about eighteen passive components, two speakers, and a single power supply with this IC. Also, it should have adequate heat sinking. This may in some cases be provided by the chassis mounting provision.

This amplifier is given a bass-boost response by the 0.02-μF capacitors in the feedback loops. As frequency decreases, the reactances of these capacitors increase; this in turn develops less negative feedback which results in greater gain at the lower frequencies. The intent is to compensate for the poor low-frequency response of small speakers.

A simple 24-volt unregulated power supply will suffice. Somewhat

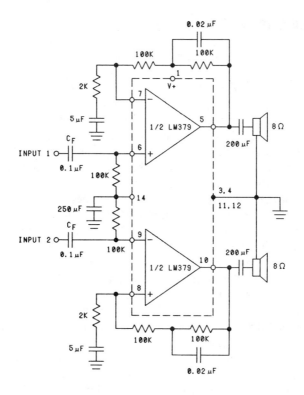

Fig. 2-8. Six watts per channel stereo amplifier using a single, dedicated power IC. (Courtesy National Semiconductor Corp.)

improved performance will result from the use of a regulated DC supply. This IC has internal current limiting and thermal protection. A nominal 70 dB of channel separation exists despite the physical proximity of the circuits on the chip. Internal compensation and about 90 dB of open-loop gain enables stable operation with the negative-feedback paths in place. It is obvious that LM379 is a dedicated IC intended for hi-fi audio applications. This will likely lead to better results than the mere adaptation of a general-purpose power IC.

The experimentally inclined may perceive this setup as a convenient means of driving higher-power output stages. In such an arrangement, a negative-feedback loop should be connected around each channel. If this is done while recording distortion, a little cut and try will lead to optimum overall performance. The easiest way to drive a more powerful output stage is via AC, that is, capacitive coupling.

BRIDGE AMPLIFIERS

Most audio amplifiers have either single-ended or push-pull output stages. Bridge-connected output stages comprising four active devices are known to have many desirable features, but the extra complication involved is generally not attractive. In other power-control applications, such as motor control and switching power supplies, the bridge format is often used, particularly where high power levels must be efficiently processed. There is, however, a low to moderate power application in audio amplifiers where the bridge configuration can be advantageously used in a simple manner.

Most power ICs have push-pull output stages. If two such ICs are used so that the single load is connected between their outputs, the load will "see" a bridge amplifier, providing the amplifier inputs are appropriately driven. Thus, the system will require only two rather than four, power devices from the hardware standpoint; even better, in some cases only a single IC will be needed because some power ICs are duals. If the power ICs are dedicated audio circuits, there will be very little effort needed to quickly assemble a quality performance amplification system because the important engineering and design techniques are already incorporated within the ICs. Incidentally, the bridge format is a neat way to use direct coupling to the speaker without the need for a split power supply or a large electrolytic coupling capacitor.

Three bridge-amplifier circuits are shown in Fig. 2-9. The common denominator in these circuits is the connection of the load or speaker between the output terminals of the power ICs. Careful inspection will show that their input connections are not the same, however. This need not lead to confusion if the basic objective of operating the output stages is kept in mind. It is simply that from the load's viewpoint, the outputs must appear to be in series-aiding, i.e., when one output is becoming increasingly positive, the other output should be, in "mirror

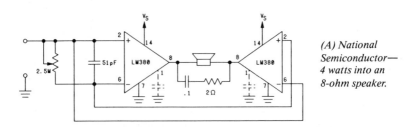

(A) National Semiconductor— 4 watts into an 8-ohm speaker.

Fig. 2-9. Three circuits, despite their differences,

fashion," becoming increasingly negative. One might very well say that the outputs are in push-pull. The fact that each output stage is already a push-pull circuit makes the resultant circuit a bridge. The load experiences twice the voltage it would be impressed with from just one of

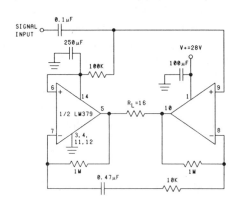

(B) National Semiconductor—
12 watts into a 16-ohm speaker.

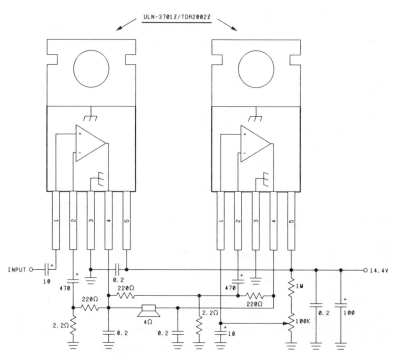

(C) Sprague Electric—115 watts into a 4-ohm speaker.

present full-bridge output stages to the load.

these push-pull output stages. It is obvious that providing the DC operating voltage is doubled, the bridge arrangement doubles available output power.

In the circuits of Fig. 2-9, note that the voltage gains of the amplifiers are set by feedback networks at 100. It is also true that the output of the left-hand amplifier is reduced potentiometrically by a factor of 100 before application to the inverting input of the right-hand amplifier. Thus both amplifiers receive equal input voltages. Phasing is taken care of by the fact that the right-hand amplifier operates as an inverter. Actually, there is not perfect symmetry between the two amplifiers; the 100K potentiometer is a balance adjustment—output current should be zeroed under quiescent conditions. The experimenter can add a volume control to the input circuit.

AUDIO POWER ICS FOR VEHICULAR USE

Power ICs for audio service in vehicles require special attention at the design and manufacturing stages. To begin with, the fact that such ICs must operate from nominal 12-volt systems aggravates the already difficult task of handling a respectable power level in a monolithic module. For a given power rating, the 12-volt IC must deal with higher currents than an IC specified for, say 20- or 30-volt operation. Worst of all is the relatively hostile environment these ICs must work in. Temperature swings are greater than that ordinarily experienced by semiconductor devices which perform in indoor equipment. Not only is the DC voltage from the battery alternator system quite variable, but it is burdened with transients that not only create interference, but are potentially destructive. Abusive conditions must be anticipated during equipment installation, battery replacement, and jump-starting. In order to merit selection over discrete-device audio circuits, the IC must provide output power in the 5- to 10-watt range, must equal or outperform the distortion and frequency response readily attained with discrete designs, and must exhibit compelling advantages in simplicity and cost. Two such ICs will be discussed.

The Sprague ULN-3703Z/TDA2003 IC power amplifier is shown in a typical circuit in Fig. 2-10. The speaker is a typical automotive type with a 3- to 4-ohm voice coil. This IC can also work into lower-impedance speaker loads. About 80 dB of open-loop gain is developed; in most applications, negative feedback results in a closed-loop gain of 40 dB. Suitable values of the feedback elements needed to bring this about

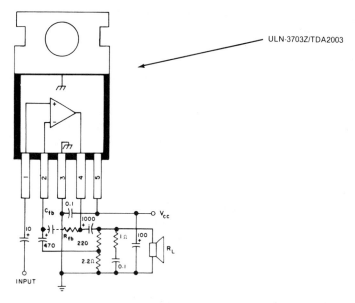

Fig. 2-10. A dedicated power IC for vehicular audio service.
(Courtesy Sprague Electric Company)

are 39 ohms for R_{fb}, and 0.039 μF, for C_{fb}. The tab of the TO-220 plastic package is at ground potential—no insulation is required for mounting to the chassis. A nominal six or seven watts are readily obtained; with a 2-ohm load, the power output can be as high as ten watts. The output devices of this IC operate essentially in class B. Total harmonic distortion does not exceed 0.15% over most of the power and frequency range. The IC is internally protected against thermal overload, output overload, supply transients, AC/DC short circuits, polarity inversion, and open grounds.

Although the circuit depicts a single speaker, operating specifications are given more realistically for two speakers in parallel, this being the format most often encountered in modern vehicles. Thus the impedance load for two 4-ohm speakers would be 2 ohms, and for two 3.2-ohm speakers it would be 1.6 ohms. Before wiring speakers, it should be ascertained that they are phased so that the speaker cones move forward and backward in unison. This precaution is more relevant for stereo than binaural systems, but attention to many little details adds up to more pleasing audio reproduction.

For those interested in higher-power capability and stereo-type delivery, two-channel audio ICs are also available. An example is the Sprague ULX-377W, a dual power amplifier of 10 watts output per channel.

Another interesting audio power IC designed for the harsh vehicular environment is the RCA CA810Q/TBA810S. This device works optimally when driving a single 4-ohm speaker, under which condition it will typically delivery six watts. It is particularly adapted to the needs of mobile communications and CB equipment. The output stage operates in class B as attested by a quiescent current drain of about 15 milliamperes for the entire IC. Because of matched transistor characteristics and accurate biasing, crossover distortion is negligible. Total harmonic distortion is in the vicinity of 0.3% for output at the 3-watt level. Open-loop gain is in the vicinity of 80 dB. Design flexibility is aided by the inordinately high input resistance of several megohms.

An audio amplifier utilizing this IC is shown in Fig. 2-11. The values of C1 and R1 in the negative-feedback loop are selected to provide a closed-loop gain of 37 dB. Two unusual pinouts on this IC are pins 6 and 9. Pin 6 is a bootstrap provision for aiding the dynamic balance in the two output transistors. Pin 9, when connected through a large capacitor to ground helps to immunize the amplifier against the effects of power supply variation. This reduces 120-Hz ripple from power-line rectifiers, and in automotive use, reduces the likelihood of alternator whine getting into the speaker.

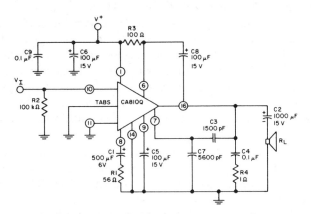

Fig. 2-11. Vehicular audio system using dedicated power IC.
(Courtesy RCA)

To provide further insight into the nature of this power IC, the schematic diagram is shown in Fig. 2-12. It is obvious that a discrete-device replica of such a circuit would be expensive, and because of the direct coupling and the high overall gain, would be tricky to balance and stabilize; it would also be difficult to simulate the quick-acting thermal feedback which protects the output stage at precise overload limits. Fig. 2-13 depicts the shutdown characteristics of this IC. The two curves reveal essentially similar information; one shows the protection in terms of output power, whereas the other curve shows the shutdown function in terms of current consumption. A nice feature of this protective technique is that normal operation automatically resumes when the abusive conditions relax sufficiently to allow the case temperature to return below 125 degrees Celsius. Note that between 125 and 150 degrees Celsius the IC remains operative, but at reduced current and power output capability. For the consumer market, such limiting action is more desirable than an abrupt turn-off action.

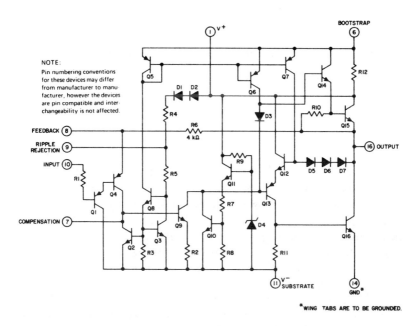

Fig. 2-12. Schematic diagram of RCA CA810Q/TBA810S audio power IC. (Courtesy RCA)

Heat removal is through two ground-potential tabs which enable the PC foil to act as a heat sink. However, free air circulation is also required in most installations. Actually, two dual-inline 16-pin packages are available—one has straight tabs, coplanar with the package; the other has right-angle tabs for convenient PC board insertion.

SOLID-STATE AMPLIFIERS / 71

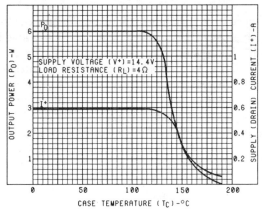

Fig. 2-13. Thermal and overload shutdown characteristics of the CA810Q audio power IC. (Courtesy RCA)

COMPLEMENTARY-SYMMETRY CHANNEL AMPLIFIER

A one- to three-watt audio amplifier can provide satisfactory volume for listening purposes. Yet, as is well known, the tendency has been to use more powerful amplifiers, up to several hundred watts. The advantage of the amplifier with greater power capability is that it exhibits a greater dynamic range because it can handle loud passages without clipping the peak waveforms. (In practice, this may be partially offset in some cases because unless well-designed, the powerful amplifier may suffer more from the effects of nonlinearity and crossover distortion at room volume.) Power is quite costly, and is generally attended by increased circuit complexity, diminished reliability, and inconveniences associated with heat removal, size, and weight. In practice, a reasonable compromise can often be realized at moderate power levels; ten to fifteen watts may indeed sound as hi-fi as much higher power levels if other performance parameters are good.

The 12-watt amplifier shown in Fig. 2-14 is relatively simple inasmuch as it utilizes an IC input amplifier, needs no driver stages or elaborate protection circuitry, and makes use of a direct-coupled complementary-symmetry output stage. Note that this amplifier is intended for use with an 8-ohm speaker. The output transistors are made specifically for service in complementary-symmetry amplifiers and are well matched in their important parameters. This yields better results than the practice of finding NPN and PNP power-transistors capable of handling the requisite power but often exhibiting appreciably different current gains, bias characteristics, and thermal parameters.

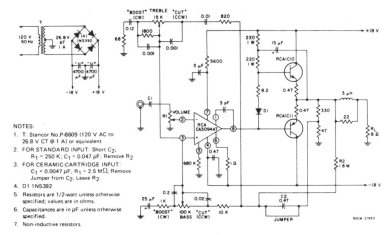

Fig. 2-14. A 12-watt complementary-symmetry audio amplifier.
(Courtesy RCA)

The integrated circuit input-amplifier facilitates implementation of treble and bass controls which are often lacking in much more powerful amplifiers. Also, special input circuit data is indicated for different input sources. In this regard, note that capacitor C1 is not merely an AC coupling capacitor, but is selected to provide appropriate frequency response. Capacitor C2 and resistor R2 are in the feedback path. They are also used to shape the frequency response to the characteristics of the source supplying the input signal. They are to be shorted or left out as indicated in the circuit notes.

The dual-polarity power supply has the topography of a bridge rectifier. However, because of the center tap on the power transformer, the supply actually operates as two separate full-wave sections, each section utilizing a pair of rectifying diodes. This has become common practice. It not only conveniently allows the dual-polarity format, but it provides full-wave rectification without the added diode voltage drop of a true bridge circuit.

A 20-WATT WORKHORSE AUDIO AMPLIFIER

The 20-watt audio amplifier shown in Fig. 2-15 is tailor-made for the experimenter and hobbyist. It uses the minimal number of discrete transistors needed to provide acceptable performance. Because of the quasi-complementary-symmetry output stage, the two power transis-

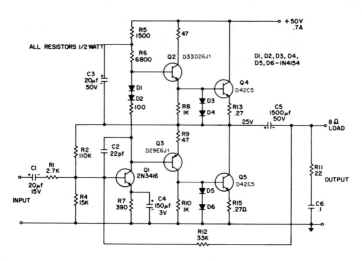

Fig. 2-15. A simple, low-cost, 20-watt audio amplifier.

tors are identical NPN types featuring plastic encapsulation in the low-cost tab-mount TO-220AB package. One single polarity 50-volt power supply with 700 milliampere capability suffices to operate this amplifier. All of the resistors can be ½-watt types, and all the diodes are identical types.

Capacitor C3 provides a small amount of positive feedback to equalize the positive and negative voltage swings at the output. This is a refinement to help make the output stage live up to its role as a symmetrical amplifier. Such localized positive feedback can be used to manipulate impedance levels and stage gains, but it does not confer any instability to the amplifier because overall feedback is always negative over the whole band of frequencies where gain exceeds unity. The main AC feedback path is through resistor R12. Overall feedback is prevented from being positive at higher-than-audio frequencies by the RC network, R11 and C6, connected across the output.

A common mistake made in the construction of quasi-complementary-symmetry amplifiers is transposition of the driver transistors, one of which is an NPN type, the other a PNP type. In this circuit, Q3 is the PNP driver transistor. Because of the way it is used, power transistor Q5 acts as if it too was a PNP type.

Output transistors Q4 and Q5 are protected from overdrive by diodes D3 through D6. Because of the selected values of R13 and R15, these diodes conduct heavily and cause a drive limiting action very soon after the development of 20 watts into an 8-ohm load.

Diodes D1 and D2 produce a voltage drop which biases the drivers, and consequently the output transistors operate into the class AB region to limit crossover distortion to an acceptable level. In this amplifier, there is adequate temperature tracking between this bias source and the bias requirements of the output transistors so that idling current of the output stage remains reasonably constant over the ambient temperature range likely to be encountered in ordinary room operation.

Because of resistor and device tolerances, it may be desirable to empirically determine the exact value of R2 so that the power supply voltage splits equally between the two output transistors. Also, R1 should not be deleted because it causes a slight attenuation of the input signal; it is used to isolate the feedback circuitry from the signal source.

40-WATT COMPLEMENTARY-SYMMETRY STEREO AMPLIFIER

The commonplace quasi-complementary-symmetry audio amplifiers naturally evoke curiosity about the real thing. Part of the popularity of the quasi circuitry goes back to an earlier era of solid-state technology when the available power device was the germanium PNP transistor. However, this popularity was enhanced by silicon technology because for a long time there were only NPN power transistors and the first silicon PNP types were inferior in ratings and reliability. That, of course is no longer the case; indeed there are many NPN-PNP pairs made specifically for complementary-symmetry output stages. Even better, there are NPN-PNP Darlington transistors also made for this purpose. Because of their relatively high current gain, Darlingtons generally dispense with two driver stages in stereo amplifier formats. This not only simplifies the circuitry and reduces cost, but helps achieve better operation and greater overall reliability.

The RCA1B07 and the RCA1B08 are NPN and PNP Darlingtons, respectively. They are specifically specified as output devices in complementary-symmetry audio amplifiers. The push-pull circuitry using these devices attains a closer match between the half-sections than can readily ensue from quasi-complementary circuits.

The block diagram of a 40-watt audio amplifier for stereo systems is shown in Fig. 2-16. This block diagram relates the various transistors to their functions in the schematic diagram of Fig. 2-17. Note the several auxiliary functions not directly involved in amplification or in power boosting, per se. These are current sources, an overload-protection circuit, and a V_{BE} multiplier. The latter circuit provides operating bias for

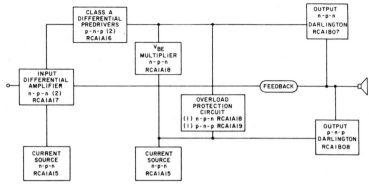

Fig. 2-16. Block diagram for a 40-watt complementary-symmetry audio amplifier. (Courtesy RCA)

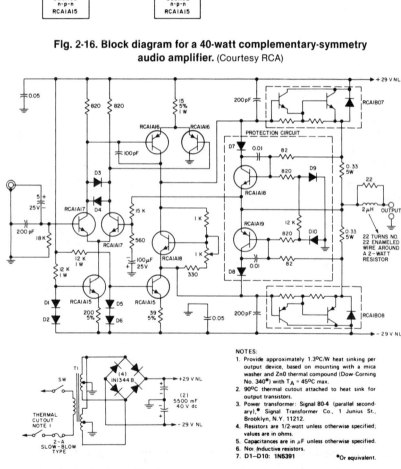

NOTES:
1. Provide approximately 1.3°C/W heat sinking per output device, based on mounting with a mica washer and ZnO thermal compound (Dow-Corning No. 340*) with T_A = 45°C max.
2. 90°C thermal cutout attached to heat sink for output transistors.
3. Power transformer: Signal 80-4 (parallel secondary),* Signal Transformer Co., 1 Junius St., Brooklyn, N.Y. 11212.
4. Resistors are 1/2-watt unless otherwise specified; values are in ohms.
5. Capacitances are in μF unless otherwise specified.
6. Non-inductive resistors.
7. D1–D10: 1N5391   *Or equivalent.

Fig. 2-17. Schematic diagram of a 40-watt complementary-symmetry audio amplifier. (Courtesy RCA)

the output stages, enabling them to maintain their adjusted class AB mode over a wide temperature range. The overload-protection circuit senses output currents in the two halves of the output stage and limits any tendency towards dangerous increases in output by reducing the drives to the Darlingtons.

The power supply does not operate as a bridge rectifier, appearance notwithstanding. Rather, it comprises two full-wave, center-tapped rectifier circuits, one for each polarity. Note that the two identical windings of the power transformer are connected in parallel.

The performance data is depicted in Table 2-1. Exceptionally wide frequency response is attained. Distortion is low because the phase and gain before feedback (that is, open-loop measurements) are quite good. It is seen that, although this amplifier is designed for 40-watts output into a 4-ohm speaker, 30 watts can be developed into an 8-ohm speaker.

Table 2-1. Typical Performance Data for a 40-Watt Complementary-Symmetry Amplifier (Courtesy RCA)

Measured at a line voltage of 120 V, $T_A = 25°C$, and a frequency of 1 kHz, unless otherwise specified.	
Power:	
Rated power (4-Ω load, at rated distortion)	40 W
Typical power (8-Ω load)	30 W
Total Harmonic Distortion:	
Rated distortion	0.5%
IM Distortion:	
10 dB below continuous power output at 60 Hz and 7 kHz (4:1)	<0.2%
IHF Power Bandwidth:	
3 dB below rated continuous power at rated distortion	5 Hz to 50 kHz
Bandwidth at 1 W	5 Hz to 100 kHz
Sensitivity:	
At continuous power-output rating	500 mV
Hum and Noise:	
Below continuous power output:	
Input shorted	100 dB
Input open	85 db
With 2 kΩ resistance on 20-ft. cable on input	97 dB
Input Resistance	18 kΩ

AUDIO AMPLIFIER WITH IC-DRIVEN OUTPUT STAGE

It was pointed out in Chapter 1 that the all discrete-device stereo amplifier can become quite complex by virtue of the many direct-coupled transistor stages required. Building and placing such amplifiers in operation can tax one's patience and pocketbook, for it is commonplace to experience heartbreaking catastrophes until all bugs are finally eliminated. Even at best, most designs are prone to oscillation problems; these often wipe out the power stage together with several of the driver and low-level circuits. Much trouble arises from physical layout, grounding, and connections inasmuch as a stereo amplifier has both high-gain and high-frequency capabilities. Adding to these inherent problems is the sloppy gain tolerances on transistors. In AC-coupled circuits this wouldn't be so difficult to cope with, but in direct coupling, it often takes a bit of cut and try to get all transistors within their proper operating and biasing ranges. As most hobbyists have learned, the less experimentation needed, the less likely there will be a horror story.

A neat solution to this dilemma is to use a dedicated IC in place of the whole assemblage of driver and low-power transistors generally found in stereo circuits. The National Semiconductor LM391 designates a series of such ICs. By using one of these very specialized ICs, it is possible to design a stereo amplifier devoid of the usual array of low-level transistors. The internal circuitry of this IC is shown in Fig. 2-18, and Fig. 2-19 depicts a 60-watt amplifier making use of only this IC and the Darlington output stage—there are no other active devices.

A further simplification could be readily made by selecting appropriate monolithic Darlington devices for the output. On a relative basis, however, much simplification has been achieved by the elimination of a half-dozen, or so, transistors. Inasmuch as these transistors are now within the LM391 IC, the stereo amplifier is electrically equivalent to traditional discrete designs, but much of it has been preengineered. This not only lifts a burden from the hobbyist, but reduces costs for the manufacturer.

Being a dedicated IC, the LM391 family yields a performance format that greatly exceeds that attainable from garden variety op amps. Drive capability embraces 10- to 100-watt stereo designs. Input noise is approximately 3 microvolts and distortion is on the order of 0.01%. The IC is internally protected for output faults and thermal overloads, and circuitry for protecting the stereo output transistors is user programma-

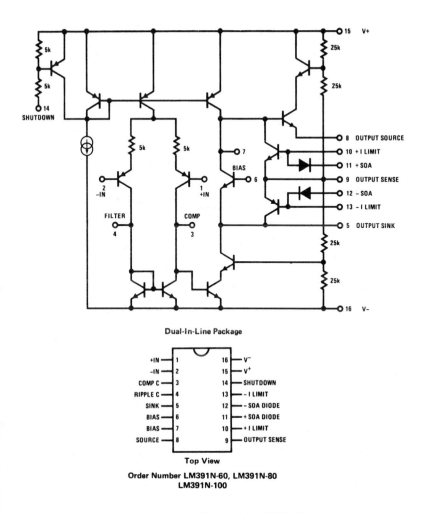

Fig. 2-18. Internal circuitry and pinout of the LM391 IC. (Courtesy National Semiconductor Corp.)

ble. Similarly, gain and bandwidth are readily selectable by the user. Further stereo protection is provided by dual-slope SOA protection and there is a shutdown pin. Rejection to the effects of supply-voltage variation is 90 dB. It is evident that the simplification provided by this IC will usually be accompanied by performance enhancement over traditional discrete designs. The pinout of the LM391 is shown by Table 2-2.

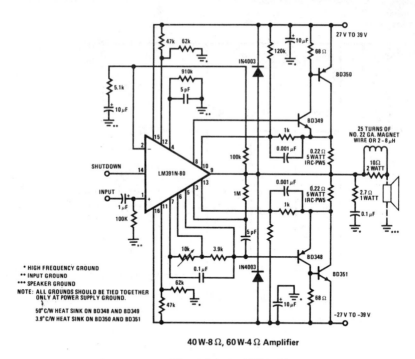

Fig. 2-19. A 60-watt amplifier using the LM391 IC. (Courtesy National Semiconductor Corp.)

120-WATT QUASI-COMPLEMENTARY-SYMMETRY STEREO AMPLIFIER

A stereo system with 120 watts per channel capability is a good compromise among a number of factors. This power level is high enough to do justice to music peaks, but sufficiently modest to hold down expenses. The overall performance of amplifiers with power ratings in this vicinity is sometimes better than is readily forthcoming from higher power systems. This is because it tends to be easier to obtain good linearity, and low hum and background noise at room volume from moderate-power amplifiers than from high-power amplifiers. The 120-watt amplifier shown in Fig. 2-20 utilizes a quasi-complementary-symmetry output stage with four identical power transistors in push-pull parallel. The quasi-complementary-symmetry circuit has a good track record in stereo practice; the operating balance between the two sections of the output stage is surprisingly good even though one section delivers power from the collectors, and the other section delivers power from the emitters of the output transistors. A worthwhile feature of this circuit is

Table 2-2. Circuit Functions of Pins on the LM391 Audio Power Driver (Courtesy National Semiconductor Corp.)

Pin No.	Pin Name	Comments
1	+Input	Audio input
2	-Input	Feedback input
3	Compensation	Sets the dominant pole
4	Ripple Filter	Improves negative supply rejection
5	Sink Output	Drives output devices and is emitter of AB bias V_{BE} multiplier
6	BIAS	Base of V_{BE} multiplier
7	BIAS	Collector of V_{BE} multiplier
8	Source Output	Drives output devices
9	Output Sense	Biases the IC and is used in protection circuits
10	+Current Limit	Base of positive side protection circuit transistor
11	+SOA Diode	Diode used for dual slope SOA protection
12	-SOA Diode	Diode used for dual slope SOA protection
13	-Current Limit	Base of negative side protection circuit transistor
14	Shutdown	Shuts off amplifier when current is pulled out of pin
15	V+	Positive supply
16	V-	Negative supply

that the output transistors are directly coupled to the 8-ohm speaker—no large electrolytic coupling capacitor is needed. This improves low-frequency response, feedback stability, and reliability.

Note the driver stages for the paralleled output transistors in Fig. 2-20. For the upper pair of paralleled RCA1B04 power transistors, the driver is a Darlington-connected RCA1C12, an NPN transistor. The driver for the lower pair of power transistors is an RCA1C13, a PNP transistor. Here, we also have a Darlington connection, but somewhat different from the more conventional Darlington circuits where both transistors are of the same type. It is this difference which causes the lower pair of RCA1804 output transistors to operate as if they were PNP, rather than NPN types. This explains the description of the output stage as having quasi-complementary-symmetry. In other words, the output stage simulates a push-pull circuit using NPN and PNP transistor types in the output stage.

Continuing the sequence of our circuit analysis from the output to the input of the amplifier, the next stage in the signal path is the differ-

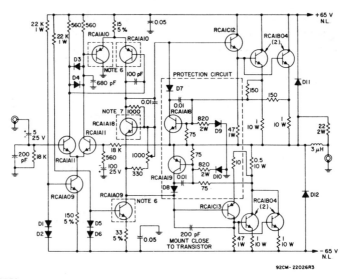

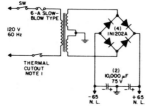

Fig. 2-20. A 120-watt quasi-complementary-symmetry audio amplifier.
(Courtesy RCA)

ential pair comprising the RCA1A10 PNP transistors. This stage is associated with a constant-current source, the RCA1A09, which is shown enclosed by a dashed-line box. The RCA1A18 transistor has a unique function and will be subsequently dealt with. Finally, we arrive at the input stage, the differential amplifier circuit comprising the RCA1A11 NPN transistor pair. This stage also operates from a constant-current source, again an RCA1A09 transistor. The incoming audio signal feeds one base of the differential input-stage; the other base of this stage is impressed with negative-feedback from the output stage.

The transistors thus far accounted for are involved in amplification of the audio signal. Additionally, there are three other transistors which serve important functions in auxiliary circuitry. One of these transistors and its associated circuitry have been referred to, but not described. This is the stage involving the RCA1A18 NPN transistor. The base-emitter voltage drop in this transistor establishes the DC operating-bias for RCA1C12 and RCA1C13 driver transistors. Because of the direct coupling between amplifier stages, this technique also provides the bias for the output transistors. Output transistor bias is a critical operating parameter in stereo amplifiers. If the output transistors are allowed to operate in pure class B, efficiency will be maximized but crossover distortion will be objectionably high. At the other extreme, class-A operation is not permissible because of the inordinately low efficiency. The best mode is a compromise—class AB, with just enough forward bias to bring crossover distortion down to acceptable levels. Forward bias beyond this impairs efficiency and causes thermal problems.

The reason the output stage bias is critical is because it is temperature-sensitive. What is needed is an adjustable bias source which temperature-tracks the output transistors. This, indeed, is the function of the RCA1A18 and associated circuitry. The two important features of this circuitry is that it operates from a constant-current source, and the RCA1A18 transistor is mounted on the heat sink of one of the output transistors. With such thermal feedback and constant current, the voltage developed across the base-emitter junction of this transistor temperature-tracks the *required* bias of the output transistors. The degree of class AB operation is adjustable via the 1000-ohm potentiometer in this circuit. If the idling current of the output stage is thereby adjusted, it will remain nearly constant over a wide range of temperature variation.

The second auxiliary circuit comprises the two transistors within the dashed box labeled protection circuit. The function of these transistors is to drastically reduce the gain of the amplifier if an attempt is made to force the output much over the rated power level of 120 watts. A study of the circuitry will show that these transistors are active loads for the second amplifier stage (the differential stage, configured around the pair of RCA1A10 PNP transistors). It will be seen that the base circuits of these protective transistors sense the voltage drops developed across the 10-watt resistors in the output circuits of the power stages. When output current tends to reach excessive levels, the protective transistors become more conductive. This depletes the gain of the second amplifier stage, and therefore of the entire amplifier. As a result, output current can only increase slightly, rather than proportionally, as would be the case without the protective circuitry.

SERVO OR MOTOR-DRIVE AMPLIFIER WITH 120-VOLT OUTPUT

A common practice in servo and motor-drive systems is to adapt an audio amplifier for the purpose. This tends to be a convenient approach because of the prevalence of low-cost powerful stereo equipment. This isn't always as straightforward as initial consideration might suggest, however. Audio amplifiers sometimes don't stand up when called upon to deliver sustained power at their advertised ratings. This is because the average power demanded from them in handling audio signals is relatively low due to the peaked waveforms involved; also, in room listening, the volume control is generally set much below that corresponding to high-power delivery. The broad frequency response is not needed, and may even prove troublesome. Usually a transformer must be procured that will provide the needed voltage for the motor—often 120-volts AC.

In light of the above problems, it may be more desirable to build an amplifier especially designed for servo systems and motor-drive purposes. The amplifier shown in Fig. 2-21 is such a circuit. The push-pull Darlington output stage delivers its power into a transformer, so that 120 volts are available for the motor. A push-pull drive signal is pro-

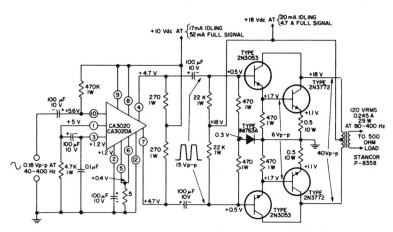

Fig. 2-21. A servo or motor-drive amplifier with a 120-volt output.
(Courtesy RCA)

vided by the CA3020 IC and no interstage coupling transformer is needed. The diode in the base return circuit of the Darlington stage breaks the path of the destructive current that might otherwise flow from avalanche breakdown in the base-emitter sections of the 2N3053 drive transistors. Note that this is not a negative-feedback amplifier inasmuch as distortion is not considered a problem in most servo and motor-drive applications. This, of course, enables operation at a higher gain than would otherwise be the case. The output stage operates class B and no detrimental effect is generally experienced from crossover dis-

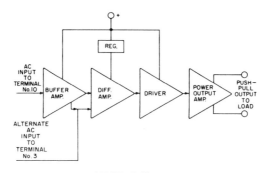

(A) Block diagram.

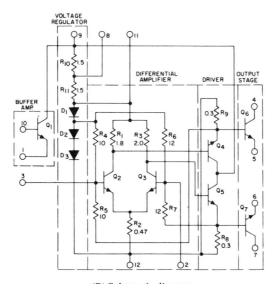

(B) Schematic diagram.

Fig. 2-22. The CA3020 integrated circuit. (Courtesy RCA)

SOLID-STATE AMPLIFIERS / 85

tortion. It can be seen that, other things being equal, fewer headaches are likely to be encountered from an amplifier of this type than from a wideband feedback amplifier of the type used in audio work.

In order to provide additional insight into this amplifier, the block diagram and schematic circuit of the CA3020 are shown in Fig. 2-22. The output transistors of this IC, Q6 and Q7, have peak current capabilities of 250 mA. This, together with the push-pull output configuration makes this IC an excellent driver for a more powerful push-pull stage.

THE RCA HC2000H OP AMP

Op amp ICs are usually small-signal devices; their traditional function has been to provide the precision processing of linear relationships between various parameters. If power is thereafter needed, it has been customary to follow the op amp with a power device.

The RCA HC2000H hybrid power module is a notable exception to this practice. This interesting device is a true op amp, having differential input, high-gain, and precisely tailored stages. But, unlike its many low-power monolithic versions, the HC2000H can deliver 100 watts RMS into a 4-ohm load, and can provide 7-ampere peak currents.

The simple amplifier circuit shown in Fig. 2-23 may fail to reveal the true nature of this device which possesses both the complexity and the

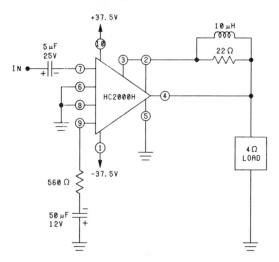

Fig. 2-23. RCA hybrid power op amp with 100-watt output. (Courtesy RCA)

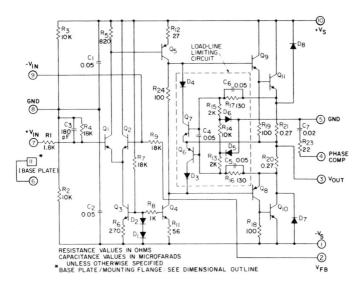

Fig. 2-24. Internal circuitry of RCA HC2000H power op amp.
(Courtesy RCA)

sophistication of its better known low-power relatives. This can be grasped from an inspection of the internal circuitry shown in Fig. 2-24. The output stage operates in class B and utilizes a quasi-complementary-symmetry configuration which permits direct coupling to the load when a split-polarity power supply is used.

Built-in protection is provided against a short-circuited load, and also to the effects of reactive loads. The metal hermetic package is both convenient and efficient for use with thermal hardware.

Looking at this device as an op amp, its open-loop gain is in the vicinity of 2000; its closed-loop gain is typically 30. (The feedback path is internal, but is brought out to a connecting wire so that both open- and closed-loop tests can be made.) In practice, the module can provide 60 watts from DC to 30-kHz under closed-loop conditions.

Designers and experimenters would do well to consider this device for applications involving audio power amplifiers, power operational amplifiers, servo amplifiers, deflection amplifiers for CRTs, solenoid drivers, linear voltage regulators, and similar power techniques. A single power supply with a DC output of 30 to 75 volts may be used in place of the split supply; in such a case, the load should then be coupled to the output (terminal 3) through a 2000-μF electrolytic capacitor, with the plus lead of the capacitor connecting to terminal 3 of the HC2000H.

SOLID–STATE AMPLIFIERS / 87

And, of course, terminal 1, which connects to the negative voltage source when using a split supply, is grounded when using a single DC supply.

RF AMPLIFIERS FOR MOBILE SERVICE

The secret of successful power-amplification at the higher frequencies is to select a transistor that has been designed for the specific type of service to be used. Although a high-frequency transistor will always work at lower frequencies, certain optimizations may be lost in this way. Other things being equal, lower-frequency transistors tend to be more electrically rugged, tend to have greater power capability, and tend to cost less per watt. By employing transistors intended for the frequency or band of frequencies needed, it is less likely that trouble will be encountered with self-oscillation or other instabilities. Additionally, if the designated RF transistor shown in a schematic diagram is used, the various reactances and resonant circuits depicted will more likely be on target than if another type of transistor is substituted. Many RF transistors have input-matching networks within their packages; these are obviously frequency-sensitive. One cannot build high-frequency amplifiers with the reckless abandon that is allowable for low-frequency projects.

Several examples of RF power amplifiers suitable for mobile operation from 12-volt systems will be described. The RF transistors used in these amplifiers are not only optimized for the frequencies involved, but they are specifically designed for operation from 12-volt storage-battery systems in vehicles. Thus, when striving for rated power output, one can be certain that the current is also within specified ratings. Not only does this prevent damage to the transistor, but it ensures that there will be no drastic drop in current or power gain as full output power is approached.

These amplifiers are for FM systems. Linear (class-B) AM service is possible. With due regard to ratings, they may also be applicable to SSB operation with suitable bias circuits. This, however, is not recommended because the transistors used have not been optimized for linearity. (Linearity requirements from a class-B amplifier are less demanding for AM than for SSB.) Note that all amplifiers have 50-ohm input and output ports.

All of the amplifiers have LC networks in their positive leads to discourage low-frequency oscillation. This is necessary because RF transistors operate on the slope of their current-gain curve; this means that their current gain increases with decreasing frequency so that at audio and low radio frequencies, the transistor is willing to oscillate with very little provocation. From a practical standpoint, this is opposite to the tendency of tube amplifiers to generate parasitic oscillations at very high frequenices, and often comes as a surprise to those whose RF experience derives from tube transmitters. Note also that no neutralization is needed for any of these amplifiers. Here again it is important to use RF power transistors with their special packaging techniques. A switching-type power transistor might possess the frequency capability for a certain RF application, but because of high internal feedback, would likely give trouble with self-oscillation.

A MOBILE 70-WATT, 50-MHz AMPLIFIER

For a solid-state amplifier at 50 MHz, 70 watts is a respectable power level, one which not too long ago was not attainable from a single consumer-oriented device. The PT8854 transistor features an inordinately low thermal impedance—only 0.86°C/W typical. It is able to efficiently transfer dissipated power to the chassis or heat sink. Making use of this RF power transistor, the amplifier shown in Fig. 2-25 develops 10 dB of power gain and requires 7 watts of RF drive. In common-emitter RF amplifiers and especially in higher-power implementations, the slightest amount of lead inductance connecting the emitter to ground will produce degeneration or negative feedback, tending to reduce power gain. This should receive top priority in establishing the layout pattern for the amplifier components. Another reason for keeping emitter inductance low is that the feedback can turn positive at some frequency, leading to instability. Keep in mind that small-diameter conductors have relatively high inductance per unit length, whereas conductors with large diameters or large cross-sections tend to have lower inductances. Wide straps are very good and are often used as conductors. In any event, the physical length of the path from emitter connection(s) to RF ground should be very short.

A MOBILE 25-WATT, 90-MHz AMPLIFIER

The RF amplifier shown in Fig. 2-26 essentially repeats the basic theme of the previous amplifier. The fact that competitive semiconductor firms are involved reveals common techniques of power boosting at

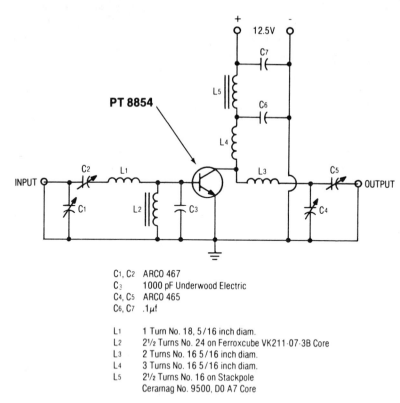

Fig. 2-25. A 70-watt, 50-MHz amplifier for mobile service. (Courtesy TRW)

high frequencies. This apparent standardization stems not so much by mutual consent, as by the harsh dictates of technology and economics. This wasn't always so; at first, power amplification at RF was more of a black art in which the experimenter or designer tried his hand at coaxing power, stability, and reproducibility from existing transistors with high f_T ratings. Usually, one objective had to be severely compromised to achieve acceptance in the others. In contrast, one can now order a dedicated RF power transistor and build an amplifier according to a rigorously designed circuit which will produce reasonably predictable results. Minimal experimentation and tweaking is necessary if construction is carried out with observance of recognized high-frequency practices.

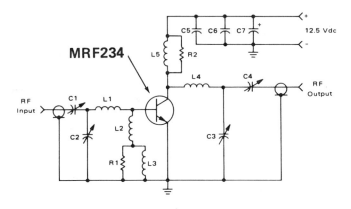

C1,C4	5.0-80 pF, ARCO 462		L3	22 μH, 9230-52 MILLER Molded Choke
C2,C3	25-280 pF, ARCO 464		L4	2 Turns, #14 AWG, 3/8" I.D., 1/4" Long
C5	1000 pF UNELCO		L5	10 Turns, #18 AWG, 1/4" I.D., wound on R2
C6	0.047 μF, ERIE disc ceramic		R1	15 Ohms, 1/2 W, 10%
C7	10 μF, 15 Vdc TANTALUM		R2	47 Ohm, 1 W Carbon
L1	1 Turn, #16 AWG, 3/8" I.D., 1/8" Long			
L2	0.22 μH, 9230-04 MILLER Molded Choke		Input/Output Connector — Type BNC	

Fig. 2-26. A 25-watt, 90-MHz amplifier for mobile service.
(Courtesy Motorola Semiconductor Products, Inc.)

A MOBILE 10-WATT, 175-MHz AMPLIFIER

VHF amplifiers may still be constructed with lumped-circuit reactances and resonant tanks. However, exceptional care must be directed towards lead lengths, spacings, and physical dimensions. At these frequencies, stray capacitance and parasitic inductance exert considerable influence on resonating frequencies, impedance matching and transformation, bypassing, and coupling. It is an in-between frequency region where stripline techniques are tantalizingly desirable, but often physically awkward. Lumped-circuit elements are satisfactory if they are physically compact and exhibit good electrical and mechanical integrity. The named brands of the components listed for this circuit represent specialists in high-frequency components; it would probably be unwise to make substitutions.

Because of antenna gain factors at both transmitting and receiving stations, a 10-watt amplifier at 175 MHz, such as the one shown in Fig. 2-27, is often capable of very reliable communications service. Antenna

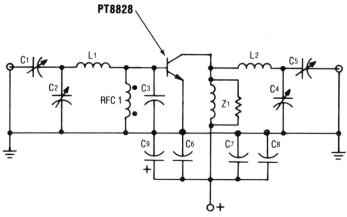

Fig. 2-27. A 10-watt, 175-MHz amplifier for mobile service. (Courtesy TRW)

PARTS LIST:
C1,2,4,5 Trimmer car, ARCO #462, 5-80pf.
C3   120pf Underwood Mfg.
C6   1000pf Underwood Mfg.
C7   0.01μf disc ceramic.
C8   0.02μf disc ceramic.
C9   25μf, electrolytic, 35 WVDC.
L1   2 T., #18 AWG., 0.25" I.D.
L2   2 T., #18 AWG., 0.25" I.D.
Z1   8 T., #18 AWG., wound on 330 ohms 1/2 W. resistor.
RFC 1   2-1/2 T., #22 AWG. on Ferrox cube VK211-17/4B Core.

gain notwithstanding, the short distances intended for coverage by many mobile services often do not justify more than 10 watts output. This circuit is relatively easy to drive, requiring approximately 750 mW of excitation.

A MOBILE 20-WATT, 470-MHz AMPLIFIER

The amplifier shown in Fig. 2-28 differs from the previous circuits in that stripline elements are used in place of lumped circuits for resonating and impedance matching. Note, also, that RF choke action is obtainable from the length of a conductor passed through a ferrite bead. Lower RF losses and better mechanical and thermal stability are more readily forthcoming from stripline reactances and resonators than from lumped circuits in this frequency range, and the power capabilities of dedicated UHF transistors can be effectively exploited. Actually, this amplifier can, by minor physical or electrical modification, be opti-

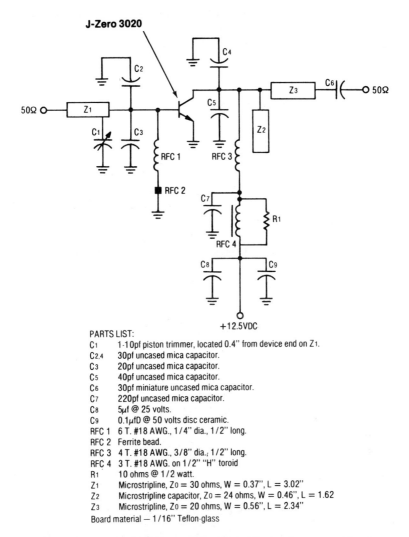

Fig. 2-28. A 20-watt, 470-MHz amplifier for mobile service. (Courtesy TRW)

mized for performance throughout the 450–512 MHz range. One might suppose that the output and frequency capability of this transistor is marginally attained; this, however, is not the case, as attested by the rated collector efficiency of 60%—this ranks with the best in RF power transistors. About 3.5 watts of input power are needed.

When constructing amplifiers with stripline elements, it is of utmost importance to use the exact board material and thickness that is specified; otherwise, resonances and reactances will not duplicate those of the original circuit. Additionally, it is possible that RF losses will be unacceptably increased.

HYBRID POWER-BOOST MODULE FOR AMATEUR 2-METER BAND

Amateur 2-meter transmitters and transceivers often operate at power levels in the fractional-watt to several-watt range. A booster amplifier with a 25-watt capability is a useful adjunct for improving communications reliability. The construction of such an amplifier can be greatly simplified, and considerable time and experimentation can be circumvented by using a dedicated hybrid module intended for this purpose. This is the Motorola MHW252 RF power amplifier module. It is broadbanded to accommodate the 144-MHz to 148-MHz range and has a minimum specified power gain of 19.2 dB. However, typical units develop sufficient gain to boost a 250-mW input to 25 watts output.

The MHW252 comprises two cascaded power stages and features thin film hybrid construction with nichrome resistive elements, berrylium-oxide substrates, and NPO, MOS, and GP dielectric capacitors. All of the devices within the module are glass-passivated. Computer-designed impedance-matching networks are used, and 50-ohm input and output impedances are maintained across the band. The internal transistors are produced on separate production lines where their parameters are optimized for the specific requirements of such a power-boost module. Thus, the manufacturer is able to specify stable performance with any load VSWR up to 6:1, at any phase angle. At the same time, output harmonics of the module itself (no output filter) are 30 dB below the output level.

From the standpoints of performance, convenience, and cost, it would be difficult to start from scratch with discrete components and equal the results that are readily forthcoming from this preengineered module. Although intended for the FM modulation format that has long been used on two meters, the enterprising experimenter may be able to obtain satisfactory results in other modes. The schematic diagram of the power booster is shown in Fig. 2-29. It will be observed that the booster is associated with an input attenuator, an output filter, and a carrier-operated relay.

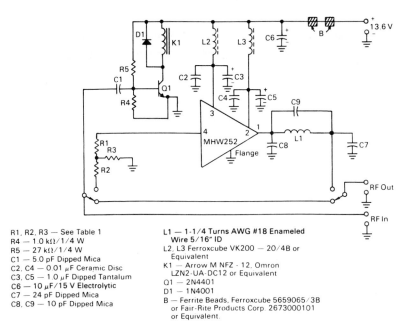

Fig. 2-29. A 25-watt power booster for the amateur 2-meter band.
(Courtesy Motorola Semiconductor Products, Inc.)

The input attenuator enables 25 watts of power to be developed for various input power levels. By its proper deployment, it protects the module against input overload when the available drive power exceeds 200 mW or so. The input attenuator can be implemented to accommodate a maximum of 5 watts of input power. In all cases, this attenuator presents 50 ohms to the drive source. Table 2-3 depicts the values of the resistors comprising the input attenuator. R1 and R2 should be 2-watt carbon types; R3 can be a 1-watt carbon resistor.

The output filter is comprised of inductor L1 and capacitors C7, C8, and C9. Although this low-pass filter has a Q of less than unity, it attenuates the second harmonic by 60 dB and the third harmonic by 50 dB. It should be noted that this filter has an M-derived configuration, that is, one of its elements is resonated. Specifically, L1 is tuned by C9 to resonate at approximately 292 MHz. Thus, in addition to the filter's normal attenuation of second- and third-harmonic energy, extra attenuation is accorded to second-harmonic frequencies of the 2-meter band. The leads of C7, C8, and C9 should be short in order to minimize the parasitic inductance of these capacitors; otherwise, the described results may not be forthcoming.

Table 2-3. Attenuator Resistances for Different Input Power Levels (Courtesy Motorola Semiconductor Products, Inc.)

The input power applied to the module must not exceed 300 milliwatts.

Power Input (Watts)	Power Ratio	Attenuation dB	R1, R2 (Ohms)	R3 (Ohms)
0.20	1.0	0	Short	Open
0.50	2.5	4	11	100
0.80	4.0	6	16	68
1.00	5.0	7	20	56
1.50	7.5	9	24	39
2.00	10.0	10	27	36
2.50	12.5	11	30	30
3.00	15.0	12	30	27
4.00	20.0	13	33	24
5.00	25.0	14	36	22

Carrier-operated relay K1 is energized when RF applied to the input of Q1 is rectified in its base-emitter diode. Note the tiny capacitor, C1, used to couple the RF signal to Q1. Q1 is forward biased to about 0.6 volt in order to increase its sensitivity. Diode D1 suppresses voltage transients which could endanger Q1.

The power module has an internal DC decoupling network which suppresses any tendency for low-frequency oscillation, down to 5 MHz. External decoupling circuitry is needed to suppress lower-frequency oscillations. Recall that RF power transistors, such as those used in the module, develop much higher gain at low frequencies than at their operating frequency. Precautions must therefore be taken against low-frequency parasitics—such oscillation can be destructive to the module. Inasmuch as this module has two stages of amplification, terminals for positive DC supply are provided for each of them. Each stage must have its individual DC decoupling network, both internally and externally. In the schematic diagram of Fig. 2-29, one of the external DC decoupling networks is comprised of L2, C2, and C3. The other is made up of L3, C4, and C5. Additionally, a mutual DC decoupling network consists of ferrite beads B in conjunction with capacitor C6. This seems to be a complicated way to provide DC power for the module. It is, however, an important procedure for proper and safe operation.

In RF work, board layout and fabrication techniques are not incidental—at VHF and higher frequencies, the orientation and spacing of parts is as important as following a schematic diagram. The PC board and parts location are shown in Fig. 2-30. For best results, these should be duplicated or at least followed in a general way.

(A) Printed circuit board pattern.

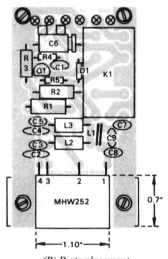

(B) Parts placement.

Fig. 2-30. Constructional guidance for 2-meter, 25-watt power booster.
(Courtesy Motorola Semiconductor Products, Inc.)

A 300-WATT, 2-30 MHz MOSFET LINEAR AMPLIFIER

300 watts PEP (peak envelope power) is considered a practical power level for reliable single-sideband communications in the amateur radio HF bands. A linear amplifier capable of providing this power with no

SOLID-STATE AMPLIFIERS / 97

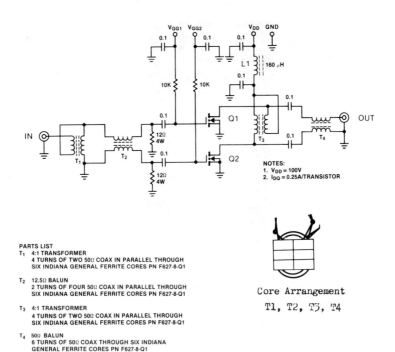

Fig. 2-31. A 300-watt, 2–30 MHz linear amplifier using push-pull RF power MOSFETs. (Courtesy Siliconix, Inc.)

tuning controls throughout the 2–30 MHz range has become an expected luxury. The MOSFET push-pull linear amplifier shown in Fig. 2-31 meets this criteria. Operating from a 100-volt DC source, the amplifier exhibits tubelike qualities in its circuit requirements. For example, the impedance levels are very manageable—they are not inordinately low as is often the case in bipolar transistor circuits. This makes it easy to implement tank circuits with practical valued elements. In this amplifier there are no resonant networks; only baluns and transformers are used in the input and output circuits.

There are other MOSFET advantages over bipolar transistor amplifiers. There is no need for the complex DC decoupling network often needed in bipolar designs. This is because the MOSFET displays no tendency to oscillate at low frequencies, its gain being substantially con-

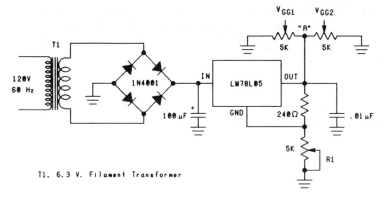

Fig. 2-32. A suitable bias supply for the 300-watt MOSFET amplifier.

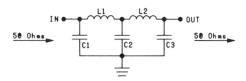

Fig. 2-33. Output filters for 300-watt MOSFET amplifier.

stant even over a broadbanded range. Also, there is no need for thermal-tracking bias circuitry. The MOSFET does not have the thermal-runaway characteristic of the bipolar transistor.

The two separate bias connections shown in the schematic diagram are intended to facilitate adjustment of equal idling currents for the two MOSFET devices. A single bias supply in conjunction with a potentiometric arrangement can be used (Fig. 2-32). Initially, R1 should be adjusted to produce about 4 volts at point A. Then, the two remaining pots can be used to provide balanced quiescent current in the two MOSFET devices.

In any event, in contrast to bipolar-transistor bias, the MOSFETs do not consume DC bias current. The idea of bias is to project the operation of the amplifier a bit into the AB region because pure class-B operation is attended by stronger harmonic generation and intermodulation products. Between 2 and 3 volts of positive bias should suffice to produce the suggested 0.25 ampere of idling current per device.

Note that the same type of core is specified for all four major core components. The inset shows the stacking arrangement of the cores for T1, T2, T3, and T4. When placed in use, a different pi-section low-pass filter must be inserted between the amplifier and the antenna for each band. This is standard practice and is needed to further attenuate harmonic energy and intermodulation products in order that these will not be radiated and cause interference. Fig. 2-33 depicts element values for suitable filters. (L1 and L2 can be empirically found by resonating with a capacitor equal to C1/2.)

The power gain over the 2–30 MHz range is a nominal $17\frac{1}{2}$ dB ± $\frac{1}{2}$ dB. Accordingly, the amplifier can be driven from a 50-ohm source with 6-watts PEP capability.

CHAPTER 3

Regulated Power Supplies

In common with traditional regulated supplies, modern circuits are of two basic types—linear and switching. Linear regulation is accomplished by means of variable dissipation. (There need not be anything actually linear in the operation of such regulators and the term is somewhat confusing. It is, however, commonly accepted in the technical literature to signify analog rather than digital circuitry.) Switching regulators operate by varying the duty cycle of a chopped wave. Newer active devices have simplified and improved the performance of both types of regulators. However, the following facts pertaining to them remain valid:

- Switching regulators are more efficient than linear types. Here, the implication of the adjective dissipative is revealing— dissipation and high efficiency are contradictory.
- Switching regulators can be packaged more compactly, particularly at higher power levels where heat sinks and blowers take up appreciable space in linear regulators.
- Above moderate power levels, say 100 watts, switchers become more cost-effective to produce and operate than linear supplies.
- For close regulation and low noise, the linear regulator is inherently superior.

The modern trend in linear regulation has been to produce all, or the greater part of the circuitry in the form of a monolithic IC. Thus, we have at our disposal brainy ICs such as the popular 723. This IC is a complete regulator in its own right, but is limited to about 100 milliamps. Used in conjunction with power devices, the overall current capability of such a simplified regulator is limited only by the rating of the external power device. Also available is a wide variety of so-called three-terminal regulators. These are power ICs generally capable of handling 1- to 10-ampere loads. Here again, external power devices can be added to greatly multiply current capability.

The modern trend in switching regulation has been to increase switching frequency. Long-used designs have operated in the vicinity of 20 kHz. Now with new and improved power devices, the switching frequency has been projected to the 50- to 200-kHz region. The advantages are smaller and lighter magnetics and filter components, easier containment of EMI and RFI, and lower production costs. A practical advantage sometimes is the avoidance of electrolytic capacitors inasmuch as much lower capacitance is needed at these higher frequencies than at 20 kHz. With traditional switchers the bipolar transistor became very dissipative at higher frequencies; however, for a price, bipolars capable of 100 kHz operation are available. Now the trend is to use the power MOSFET in high-frequency switching supplies. Yet another important development for switchers has been availability of dedicated ICs for generating the clock frequency and processing the pulse-width modulation waveforms.

SIMPLE, USEFUL SINGLE TRANSISTOR POWER-SUPPLY CIRCUITS

Power transistors and power Darlingtons can perform useful power-supply functions in very simply circuits. Fig. 3-1 depicts four configurations that have almost universal application. Because of the variable requirements of individual projects, parts values are not given. However, sufficient guidance will be provided to enable the knowledgeable experimenter to implement these circuits. All are quite tolerant to operating conditions. All may be implemented around either NPN or PNP power transistors. If the opposite-polarity transistor is used from that indicated in Fig. 3-1, just reverse DC polarities, zener-diode polarities, and electrolytic-capacitor polarities. Also, these circuits will tend to work even better with power Darlingtons if allowance is made for the fact that the voltage drop and power dissipation of Darlingtons will often be higher than that of single transistors.

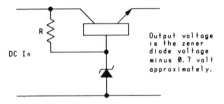

(A) Series-pass or emitter-follower voltage regulator.

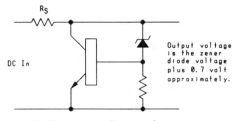

(B) Shunt-type voltage regulator.

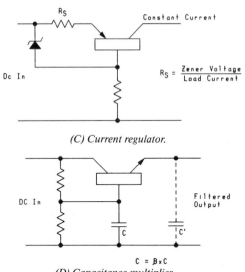

(C) Current regulator.

(D) Capacitance multiplier.

Fig. 3-1. Simple single-transistor circuits for power supply applications.

The circuit of Fig. 3-1A is that of a series-pass voltage regulator. It is basically an emitter-follower with the base-bias voltage stabilized by a zener diode. The output voltage is the zener-diode voltage minus the base-emitter voltage of the transistor. For many applications, a 1-watt zener-diode with breakdown voltage in the 6- to 10-volt range works out well. R can then be chosen to allow 10 to 50 milliamps of zener-diode current for a start. Exact design depends upon the DC input voltage, the current gain of the transistor, and the load current.

The circuit shown in Fig. 3-1B is also a voltage regulator, but here the power transistor acts as a variable shunt resistance. Voltage regulation occurs as the transistor varies its internal resistance with respect to series resistance R_S. This type of voltage regulator is very efficient for a constant load which is heavy enough to *almost* deplete current flow through the transistor. For lighter loads, the power dissipation in the transistor is high and the efficiency is low. Note that this characteristic of the shunt voltage regulator is just opposite to that of the previous regulator circuit which becomes progressively *less* efficient as loading is increased. Series resistance R_S is selected to enable regulation to prevail for, say, 110% or 115% of the expected load current. This is intended as a margin to allow for vagaries in the DC input voltage from the rectifier filter system. In any event, such an empirical procedure will provide a good start for further optimization. The zener-diode circuit can be treated in the same way as in the previous regulator.

Via another permutation of the several components involved, we arrive at the current regulator of Fig. 3-1C. By appropriate selection of the zener voltage and the value of R_S, a constant current will flow through a wide range of loads. This constant current will prevail for loads from short-circuit (zero resistance) up to some maximum resistance. The product of load resistance and load current cannot exceed the DC input voltage. Indeed, this product has to be somewhat less than the DC input voltage because of the voltage drops in R_S and in the transistor itself. Other things being equal, the higher the DC input voltage, the higher the load resistance that can be accommodated. Constant-current regulators of this type find applications with electromagnetic devices and motors, for electroplating and battery charging, and for electronic purposes, such as the simulation of high impedance, or the linear charging of capacitors. The constant-current mode is an excellent way to operate filaments in various devices, such as transmitting tubes and projection lamps. This is because the thermionic emissivity, the optical output, and the life span of filaments are more directly related to current than to voltage.

The interesting circuit shown in Fig. 3-1D is somewhat similar to that of Fig. 3-1A. However, a capacitor has been substituted for the zener-diode voltage reference. As might be expected, this circuit no longer has the ability to behave as a DC voltage regulator. However, it now exhibits very useful AC properties. An amplified capacitance, C', now appears across the output terminals and tends to be more cost effective in filtering ripple than an equivalent physical capacitor of that size. Actually, the capacitance of the small physical capacitor, C, is multiplied by the current gain, B, of the transistor. Power Darlingtons, because of their high B values, are particularly effective in this circuit. The low output impedance to AC provided by this circuit makes it useful in stereo and TV applications where common-impedance coupling between various stages and sections must be avoided. The base-biasing resistances are chosen to produce the lowest voltage drop across the transistor consistent with the desired capacitance multiplying action. A thousand ohms for each of these resistances is a good starting point for experimentation.

It should not be supposed that the multiplied capacitance, C', has DC energy-storage properties. It is real only in its effect on AC. Manufacturers of consumer products sometimes realize worthwhile savings in both cost and bulk by using this circuit as an electronic filter in conjunction with a poorly filtered half-wave rectifier circuit.

SINGLE TRANSISTOR VOLTAGE REGULATORS

The circuits shown in Fig. 3-2 are shunt-type voltage regulators which operate in similar fashion to the shunt-type voltage regulator of Fig. 3-1B. However, in these circuits, a low-voltage zener diode is able to regulate a higher voltage. In principle, any voltage can be thus regulated if the voltage-reduction ratio provided by divider network R1, R2 is appropriate. Of course, it is also true that the shunt transistor must be selected with an adequate voltage rating. The NPN and PNP circuits of Fig. 3-2A and Fig. 3-2B, respectively, work in essentially the same way. In previous years, not many PNP silicon power transistors were available, but now one has about the same choice of voltage and current capabilities with either type.

In implementing these circuits, the current drawn by voltage-divider network R1, R2, should be at least five times the zener-diode current. R_s should be as high as is consistent with the input voltage and load current involved. By making R_s high, minimal dissipation takes place in the transistor, thus optimizing the efficiency of the regulator. This statement, however, assumes that a fixed load is used. If there will be variation in load current, R_s must be made lower in order to accommodate the heavier loads. As previously pointed out, shunt regulators are most efficient when tailored to regulate for a fixed load.

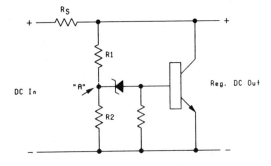

(A) High-voltage shunt regulator using an NPN transistor.

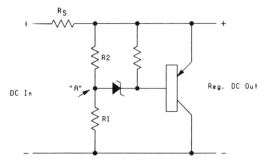

(B) High-voltage shunt regulator using a PNP transistor.

Fig. 3-2. Single transistor voltage regulator circuits.

These circuits operate in such a manner as to maintain circuit junction A at a potential which is the sum of the zener-diode voltage and the base-emitter voltage of the transistor. Thus, the output voltage is $[(R2/(R1 + R2)](V_z) + V_{BE}$. Inasmuch as the best zener diodes have breakdown voltages approximately in the 6- to 10-volt region, these circuits provide a convenient means of obtaining reasonably good voltage regulation in the several-hundred volt region. Much, of course, depends upon the integrity of the divider resistors. Generally these should be wirewound types, or precision resistors with low temperature coefficients. In many applications, circuit B will prove advantageous over circuit A in that the transistor need not be electrically insulated from the heat sink. Whether this merits consideration depends upon the power levels involved.

There are two nice features about shunt-type regulators that are often overlooked. First, these regulators draw constant current from the DC input source, regardless of load current. This tends to simplify design and implementation, as well as reduce the cost of the unregulated supply. Second, the shunt-type regulator is short-circuit proof insofar as concerns the safety of the regulating transistor. Various versions of these circuits are found in TV sets, where they are often used with Darlington transistors.

A TWO-TERMINAL CONSTANT-CURRENT DIODE

Constant current is required for various purposes. A common way of achieving this is to make a simple conversion of a voltage-regulated power supply. Instead of sensing output voltage, the error amplifier of the regulator is made to sense the voltage drop across a small resistance placed in series with the load. This arrangement can, indeed, produce very close current regulation, or can provide a preset constant current. However, it often would be more convenient to deal with a two-terminal current regulator, rather than the three-terminal (minimum) format that follows from converting the voltage regulator to a current regulator.

A number of two-terminal constant-current diodes have been on the market for quite some time; these provide constant currents in the microampere and milliampere ranges, usually being limited to currents below several tens of milliamperes. Actually, these devices are not diodes, but use either one or two JFETS in the simple circuits. (Older versions used bipolar transistors.)

Naturally, it would be nice to be able to provide constant currents in the ampere range using such a two-terminal analog of the zener diode. This can be done by using a power MOSFET in conjunction with a garden-variety op amp. The circuit shown in Fig. 3-3 can provide currents from about 10 milliamperes to 12.5 amperes by adjusting variable resistance R1. Note that unlike a conventional current-regulated power supply, there is no ground or return connection—rather, the two terminals of the circuit are simply inserted in the "hot" side of the load. Thus, the entire circuit behaves as a smart resistor; that is, one which automatically keeps current through itself at a constant value. This automatic-regulating feature prevails right down to a short-circuit load.

The current source supplying I_i in Fig. 3-3 must have a voltage of at least 15 volts with respect to ground. The cold side of the load connects to ground. Therefore, the load receives its current through the source-drain section of the power MOSFET. Interestingly, this circuit also utilizes one of the low-current diodes referred to above. This diode, in

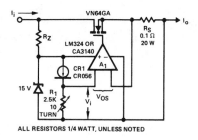

Fig. 3-3. Two-terminal constant-current source using a power MOSFET.
(Courtesy Siliconix, Inc.)

conjunction with variable resistor R1, forms a variable-voltage, simulated-zener reference source. Coincidentally, the + and − designations on the op amp not only indicate the polarity of the DC operating voltage, but also the noninverting and inverting signal-input terminals, respectively.

QUICK AND EASY POLARITY INVERSION

Sometimes a low-power DC source with reversed polarity from that of the main supply is needed. One way of producing such a voltage is to construct a small inverter using various transformer circuits. The output of such an inverter can then be rectified and filtered to provide the desired voltage. Tight regulation of such sources is not usually called for. They are used for such purposes as biasing substrates of certain ICs, biasing switching-transistors to help sweep out stored charge during off periods (thereby increasing switching speed), and they are sometimes used for logic devices.

There is another technique for generating a reverse-polarity voltage which does not require any magnetic-core components, and which is very simple and inexpensive to implement. This involves a so-called "charge pump," which is a circuit for alternately charging and extracting charge from a capacitor. By means of a steering diode, the extracted charge polarity can be selected to be *opposite* to that of the charging source (the main supply). The circuit of a practical charge pump is shown in Fig. 3-4. Capacitor C is alternately charged and discharged by means of the MOSFET switch, which is driven at a 100-kHz rate by the self-excited pulse oscillator configured around the 74C14 inverter. The output of the charge pump is maintained at 5 volts by the shunt-connected zener diode.

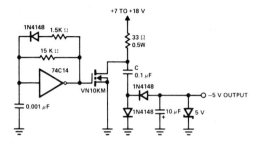

Fig. 3-4. A way of obtaining a negative voltage from a positive DC source. (Courtesy Siliconix, Inc.)

Optimum operation does not occur with a 50% duty cycle switching wave from the pulse generator and a little experimentation may be in order to properly supply a given load. The 1500-ohm resistor and the 15,000-ohm resistor in the feedback path of the inverter can be slightly modified in order to achieve optimum operation. This can be done by inserting a milliammeter in series with the zener diode—the idea is to find the duty cycle of the switching wave which delivers the highest current into the zener diode. In varying the duty cycle keep in mind that, other things being equal, the higher the switching rate the better the results. Both the 74C14 inverter and the VN10KM MOSFET perform well in the vicinity of 100 kHz. For equivalent results at lower frequencies, the values of capacitor C and the filter capacitor would have to be increased.

Don't neglect to connect the 74C14 to a DC operating voltage. (Such connections are generally not shown in logic-circuit diagrams.) Note that the conventional current-limiting resistance used with zener diodes is not needed in this circuit—the loading effect of the −5-volt output circuit suffices for this function.

TRACKING REGULATOR USING THREE ACTIVE DEVICES

The three-terminal IC regulator has understandably made a strong impact on power supply technology for the simple reason that these power ICs are self-contained regulated power supplies. Moreover, it is almost a matter of regarding such a subsystem as just another circuit component. Actually, successful implementation often requires some street-wisdom, as is delineated in Chapter 1. By observing the hints, kinks, and precautions therein covered, it becomes essentially a matter of selecting the right regulator for the job. Therefore, it would serve no useful purpose to deal with representative power supplies using these ICs—the material in Chapter 1 more than suffices.

In addition to the direct and simple use of the three-terminal regulator by itself, various associations with other active devices often yield profitable results. For example, the tracking regulator shown in Fig. 3-5 makes use of three different active devices—a three-terminal regulator IC, an op amp, and a series-pass transistor. The negative output voltage tracks the positive output voltage, and both are approximately equally well regulated. Providing the unregulated power supply is adequate, this dual supply can deliver up to 5 amperes from each regulated output.

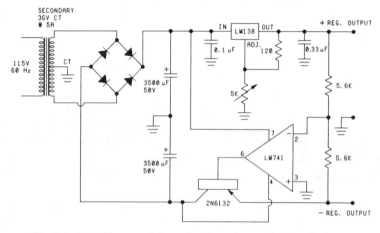

Fig. 3-5. A tracking regulator using three types of active devices.

The upper half of the schematic diagram is conventional for the LM138. The lower half reveals an op amp nulling circuit in conjunction with a PNP series-pass transistor. Note that the op amp senses the midpoint of the positive and negative output voltages. If the midpoint is *positive* with respect to ground (positive output higher than negative output) the op amp tends to close down the series-pass transistor, thereby increasing the negative output voltage. If the op amp senses a *negative* voltage at the midpoint (negative voltage higher than the positive output voltage), the op amp tends to open up the series-pass transistor, thereby decreasing the negative output voltage. In both of these corrective actions, the op amp ceases its corrective action when the midpoint of the 5.6K resistors is at zero potential with respect to ground.

This condition corresponds to the desired equality between the positive and negative output voltages.

An interesting and useful aspect of this circuit is that the tracking accuracy remains good even for unbalanced loads. However, much depends upon the integrity of the 5.6K resistors. Rather than use composition types, it is better to use low temperature coefficient metallic-oxide resistors with 1% tolerance. Otherwise, this circuit is both flexible and forgiving; a wide variety of active devices other than the designated types can be made to work satisfactorily. If a grounded heat-sink is used for the series-pass transistor, a mica or ceramic insulating spacer will be needed.

SYNCHRONOUS RECTIFICATION

Synchronous rectification has a long history, but in the past it rarely offered unique advantages for electronics applications. Recently, this technique has been revived because higher rectification efficiencies can sometimes be attained than by the use of either ordinary or exotic rectifier diodes. Synchronous rectification occurs when the control electrode of a three-terminal active device is impressed with an appropriately timed turn-on signal. The simple circuit employing a PNP germanium transistor shown in Fig. 3-6A provides quick insight into the basic concept. The selection of the old-fashioned power transistor will be explained subsequently.

In the circuit of Fig. 3-6A, the arrangement is such that the transistor receives forward bias at the same time the emitter-collector circuit is properly polarized for collector current, i.e., when the collector is negative with respect to the emitter. Thus, the transistor is synchronously turned on to supply current to the load for one-half of the AC cycle. This constitutes half-wave rectification. This simple circuit is not very practical, primarily because the transistor is not likely to be saturated over the full duration of the rectified wave. Next to the synchronization process, saturated operation of the transistor is very important; it brings about high rectification efficiency, and protects the transistor from destructive dissipation.

Fig. 3-6B shows a more practical implementation. This circuit operates with a steep-sided AC wave, which can be a square wave, a rectangular or stepped wave, or a PWM wavetrain. The small signal diode in the base lead prevents the emitter-base diode from being driven into the reverse-breakdown region; this diode also tends to promote quicker

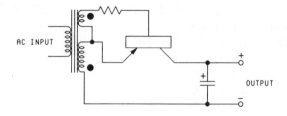

(A) Basic circuit for demonstration.

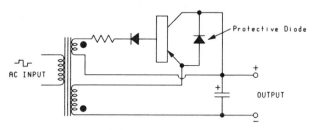

(B) Practical circuit for high-current, low-voltage loads.

Fig. 3-6. Germanium power transistor as a synchronous rectifier.

turn-off of the transistor. The diode shown connected from emitter to collector protects the transistor if it should come out of saturation. This could be caused by inadequate AC drive, or by some load conditions. Note that a choke is not included for filtering—both inductive and excessive capacitive loading can cause penetration of the transistor's SOA rating. Under normal load conditions, satisfactory operation can usually be had from just enough capacitive filtering to reduce ripple sufficiently.

The reason for specifying the PNP germanium transistor is that this almost forgotten device has an exceedingly low value of $V_{C(sat)}$. Indeed, its forward voltage drop can be much less than that of the best silicon-junction rectifier diode, and can be comparable to, or better than, that of Schottky diodes. This implementation is intended for low voltage—5 volts or less—power supplies which deliver high current, say 10 to 50 amperes. Inasmuch as germanium transistors are low-frequency devices, their application must be restricted to perhaps several kilohertz. Where applicable, however, such synchronous rectifiers may merit consideration in terms of both cost and operating efficiency. When the rectifier is initially placed in operation, the transistor remains out of saturation while the filter capacitor is charging. During this time, the protective diode is forward biased and absorbs much of the dissipation that would otherwise have to be handled by the transistor. When the capacitor is nearly fully charged, the transistor is able to go into saturation, at which time the voltage across the protective diode becomes too low for conduction. Therefore, the protective diode remains inactive unless some load or circuit disturbance again causes the transistor to fall out of saturation. This mode of operation is possible because the protective diode needs about six or seven hundred millivolts to conduct, whereas the voltage applied to it when the germanium transistor is saturated tends to be in the vicinity of 100 to 300 millivolts.

The major surviving manufacturer of germanium transistors is Germanium Power Devices Corp. This firm makes devices which are optimized for deployment as synchronous rectifiers, some of which have 100-ampere ratings.

The synchronous rectifier scheme of Fig. 3-7 provides full-wave rectification from silicon NPN power transistors. While $V_{C(sat)}$ is not as low as that attainable from germanium power transistors, the forward voltage drops can, nevertheless, be appreciably lower than that of silicon rectifier diodes. Silicon transistors have higher frequency capabilities than germanium transistors and, of course, have thermal advantages. However, manufacturers should be queried for special power transistors intended for use in synchronous rectifier circuits. The circuit of Fig. 3-7 operates successfully with an LC filter. The secondary winding on the filter choke generates voltage pulses which help turn off the transistors. C1 is a speedup capacitor for this function, and R1, R2, and R3 are current-limiting resistors. Diode D2 is the free-wheeling or catch diode commonly used in switching supplies and inverters. It is not as important in a full-wave configuration as it is in a half-wave configuration, but in practical systems it is still needed to transfer to the load the energy stored in the choke. D2 and D3 are small-signal diodes which help steer base signals between the two transistors. The Unitrode Corp. makes special silicon power-transistors optimized for synchronous rectifier operation.

POWER MOSFETS IN A SYNCHRONOUS RECTIFIER BRIDGE

The use of power MOSFETs in synchronous rectifiers stems from several considerations. The fact that these devices have frequency capabilities comparable to Schottky diodes is initially an attractive feature. The fact that they can handle higher voltages is important too, inasmuch as this is one of the drawbacks of the Schottky device. Until recently, however, the drain-source resistance of the power MOSFET has

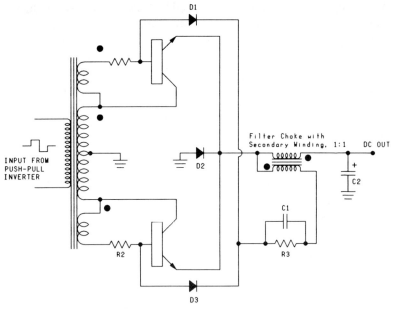

Fig. 3-7. Synchronous rectifier system for push-pull inverters.

been relatively high and discouraged consideration of the device as a rectifier. However, in many applications this no longer looms as a problem and there is an ongoing trend for reduction of the resistance—what was once assumed to be an intrinsic shortcoming of this device is no longer viewed that way. Newer power MOSFETs can perform very efficiently as rectifiers because of their high-frequency response and their acceptable on resistance. Thus, they are prime candidates for switching power-supply designs, especially those involving switching rates beyond 20 kHz.

Another compelling factor for considering the use of power MOSFETs in synchronous rectifiers is that they have become available in P- as well as N-channel types. This facilitates circuit implementation. Also, power MOSFETs have become cost-competitive with other rectifiers and rectification techniques. Of even greater importance, semiconductor companies are optimizing certain power MOSFETs for synchronous rectifier service.

The bridge-configured synchronous rectifier shown in Fig. 3-8 is interesting because of its simplicity. Note that the rectifying elements comprise two N-channel and two P-channel devices. In tracing out the operating cycle of this scheme, it should be kept in mind that power

MOSFETs, unlike bipolar transistors, can conduct with reversed output-circuit polarity. Thus, a P-channel MOSFET, which ordinarily operates with the drain negative with respect to the source, can also operate with the drain positive with respect to the source.

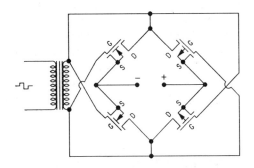

Fig. 3-8. Power MOSFET bridge circuit for synchronous rectification.

It is important to use devices with low values of R_{DS} in this circuit. This is not merely to obtain high rectification efficiency. If R_{DS} is too high, the voltage drop would then be sufficient to turn on the parasitic diode which all power MOSFETs contain between drain and source. See Fig. 3-9 if this happens, the circuit of Fig. 3-8 will still perform as a full-wave bridge rectifier. However, the MOSFETs, as such, will now be out of the picture and rectification will be performed by the parasitic diodes. But such a conventional rectifier will exhibit both high conductive losses and high switching losses. An easy check to determine the

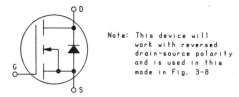

(A) Realistic symbol of an N-channel power MOSFET.

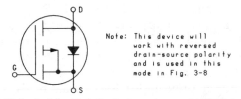

(B) Realistic symbol of a P-channel power MOSFET

Fig. 3-9. Power MOSFET symbols showing parasitic diodes.

mode of rectification is to measure the drain-source voltage drops. If they are less than 1 volt, it is probable that synchronous rectification is taking place. Much higher voltage drops than this infer at least some undesired forward conduction of the parasitic diodes.

As in the synchronous rectifiers using bipolar transistors, the manufacturers of power MOSFETs should be consulted for information pertaining to power MOSFETs optimized for use as synchronous rectifiers. Such devices will have inordinately low values of B_{DS}. They will thereby provide high rectification efficiency because of a low forward voltage drop. Generally, such dedicated devices also display gate turn-on voltages lower than those intended for use in conventional amplifier applications.

POWER MOSFET IN ELECTRONICALLY CONTROLLED RECTIFIER

An ideal rectifier conducts perfectly in one direction of current flow, but totally impedes flow in the opposite direction. Practical semiconductor rectifiers require a threshold voltage to initiate current flow in the forward direction, and thereafter develop an increased voltage drop because of ohmic resistance. In the reverse direction, they may or may not limit current to a satisfactory degree—much depends on material, fabrication, and operating temperature. There are other nonideal features of these rectifying diodes; their rectifying efficiencies may show considerable degradation as the frequency is raised. Or, the reverse polarity blocking ability may not hold at the voltage levels being used because of zener or avalanche breakdown. And sometimes capacitance, nonlinearity, and noise production enter the picture. One aspect of junction-diode behavior—forward voltage drop—merits particular attention in power-control circuits because of the power loss that is thereby incurred.

Several ways of dealing with forward voltage drop have evolved. In principle, the so-called contact voltage of the PN junction can be reduced by several orders of magnitude by means of an operational half-wave-rectifying circuit. Here, the 0.6 volt or so across the junction is divided by the open-loop gain of the op amp, and the overall circuit performs much like a point-contact diode with a near-zero threshold voltage. This approach is fine for low-level signals, but is not practical for the voltages and currents found in power work. A second approach involves the selection of semiconductor materials. Paradoxically, the best material from the standpoint of low-threshold voltage is the once popular germanium. To this day, the favorite advertising theme of one of the few remaining germanium semiconductor firms is the fact that large germanium rectifying diodes can perform with less voltage drop than either silicon PN diodes or silicon Schottky diodes. Unfortunately, germanium devices have limitations in other characteristics; their thermal characteristics are inferior to those of silicon devices. And germanium transistors, which show very good rectifying efficiency in low-frequency synchronous rectifying circuits, do not display the frequency capability of silicon transistors.

Schottky diodes rectify well at high frequencies and develop appreciably less forward voltage drop than do silicon diodes made with conventional junctions. Although the forward conduction behavior of Schottky diodes withstands high temperatures well, the reverse conduction at high temperatures is as troublesome as it is with germanium. Moreover, the reverse conduction degrades rapidly as a function of voltage. Although this may not always be so, Schottky diodes have a track record of being relatively costly components.

Gallium arsenide devices show much promise for simultaneous operation at high temperatures and at high frequencies, but diodes and transistors made of this material have much higher voltage drops than do silicon devices. (An example is the nominal 1.7-volt drop across an LED—basically a gallium arsenide PN diode.) Nor can we go back to earlier-used materials such as selenium or copper oxide. Such materials had unique features, to be sure. Generally, they were quite leaky rectifiers, i.e., they displayed poor forward-to-reverse conduction ratio. Moreover, these earlier materials tended to lose their rectifying properties with moderate temperature rises.

From the above, it should be evident that the subject of rectification remains an important theme in power engineering. It is true that in many cases, such as rectifying from the 60-Hz power line, we can just throw in some garden-variety silicon rectifiers and our troubles are over. But in modern power technology involving frequencies into the RF region, square waves, and high-speed switching processes, improvements in rectification are very much needed. Agitating this need has been the move to 5-volt heavy-current power requirements for logic and computer equipment.

A recent contribution to rectification technology involves the use of power MOSFETs. Hitherto, these devices have been used in synchronous-rectifier circuits and manufacturers have been optimizing MOSFETs for this dedicated purpose. A somewhat different circuit approach is shown in Fig. 3-10. This circuit is suggestive of synchronous rectification in the sense that a three-element power device is made to rectify by application of an appropriate signal to its control element. The circuit is also suggestive of precision half-wave rectifiers using op amps in the use of an op amp to participate in the rectification process. Yet, the circuit is unique in that its operation differs from both of the aforementioned rectification techniques.

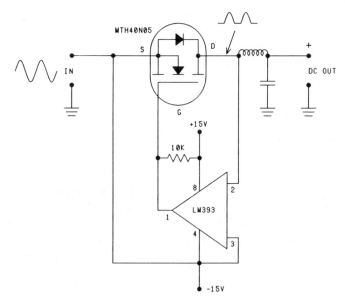

Fig. 3-10. Efficient rectification with a power MOSFET.
(Courtesy Motorola Semiconductor Products, Inc.)

The op amp depicted in Fig. 3-10 is more specifically a voltage comparator. It functions in this circuit in such a way that whenever the input voltage is more positive than the output voltage, the gate of the power MOSFET receives forward bias and conduction takes place in this device. Otherwise, the gate is biased off and no conduction occurs in the MOSFET. Thus, the MOSFET is gated to perform as a unidirectional device, or rectifier. A natural question regarding this circuit performance pertains to the intrinsic or parasitic diode of the power MOSFET—it would appear that this diode would short out the source-drain circuit, thereby rendering the MOSFET useless in the circuit. The reason this does not occur is not obvious from the topography of the circuit; it is because the voltage drop from source to drain is much less than the ordinary contact potential (about 0.65 volt) of the parasitic diode. Therefore, the parasitic diode is not given the opportunity to become active.

Inasmuch as power MOSFETs have established a respectable record as power amplifiers and power switches, one might well ponder why rectifying circuits such as this one did not come into prominence. The reason is that power MOSFETs were long plagued with high ON resistances. In order for the MOSFET to successfully operate in this type of circuit, the ON resistance must be low enough so that the parasitic diode will not conduct. Forward conductance of the parasitic diode is not necessarily damaging to the device, nor does it prevent the intended half-wave rectification of the circuit. However, by the time the parasitic diode conducts, the circuit has lost its principal advantages over other rectifier schemes—low forward voltage drop, and high rectification efficiency. Recently, Motorola and other manufacturers have made available power MOSFETs with both lower ON resistances and greater current-handling capability than previous devices. As an example, the Motorola MTH4ON05 TMOS device used in this circuit has an ON resistance of only 0.028 ohm when delivering 10 amperes of rectified DC. Thus, the voltage drop is under 300 millivolts—better than can be achieved with a Schottky diode.

As shown, a ± 15-volt low-current auxiliary source is needed for the circuit. Once the experimenter becomes cognizant of the circuit's basic principle, many different things can be done to tailor performance to individual project requirements. For example, the high-frequency performance of this circuit, as depicted, would probably be limited by the op amp long before the frequency response of the power MOSFET becomes a factor. Also, even though rectification from a sine-wave source is illustrated, operation from a square-wave source would increase efficiency and relax filtering requirements.

When evaluating this circuit, ascertain that the voltage drop across the source-drain terminals of the power MOSFET is low, as described. If the op amp circuit is not working properly for some reason, the overall circuit could still behave as a good rectifier by virtue of the parasitic diode taking over this function. Of course, the extra margin of performance possible with intended participation of the MOSFET would not be forthcoming.

Builders of power supplies such as this one often suppose that the greatest stress on the series-pass transistor occurs at half load, full load, or in any event, somewhere in the normal operating region. Actually the power dissipation in the series-pass transistor maximizes in the *foldback* region. As a rule of thumb, this tends to occur when the output voltage has fallen to about two-thirds its normal (regulated) value. That is why a liberally sized heat-sink is mandatory.

Foldback action is initiated when op amp IC2 monitors a sufficient voltage drop across sensing resistance R_S to turn on transistor Q3. When this happens, the voltage developed across resistance R_C acts at terminal 1 of the CA3055 IC voltage regulator to reduce its output current. This, in turn, reduces the drive of transistor Q2 and thereby of series-pass transistor Q1. Thus, the output current of the supply is decreased as the load becomes heavier. Note that this action overrides the normal regulating action. For a full short circuit, the load current approaches, but does not quite attain, zero. In this supply, this residual load current is about 120 milliamperes.

The fixed portion of current-sensing resistance R_S should be approximately 0.064 ohm. This could be implemented from a length of copper wire, but because of the effect of temperature, it is much better to use manganin wire. In this regard, Driver-Harris #18 manganin wire has a resistance of 0.176 ohm per foot and exhibits a very low temperature coefficient.

It will be noted that potentiometer R20 is also involved in the current-sensing circuit. This is intended as a one-time factory-adjust provision. With R_S equal to zero (shorted out by a short length of heavy conductor) adjust R20 so that there are 200 millivolts between its wiper and ground. This adjustment is made under the condition of zero load current. Next, remove the shorting conductor and set the variable portion of R_S at its zero ohms position. The idea now is to determine the *exact* length of resistance wire in the fixed portion of R_S to allow onset of foldback operation at a load current of 3.15 amperes. A stable load and accurate current meter are mandatory for this adjustment. Thereafter, the variable portion of R_S will enable panel adjustment of the onset of foldback operation at load currents below 3.15 amperes.

The deserved popularity of the 2N3055 power transistor has been alluded to; it would only be fair to also call attention to the fact that standardization of this transistor hasn't been as rigorous as it might have been—there have been considerable differences among brands, and even among production runs of the same brand. RCA, however, prides itself on the close supervision and tight quality control of this device.

Q4 and associated circuitry form a shunt-regulated auxiliary supply for IC2 and Q3 of the foldback control section. Zener-diode CR5 protects IC1 from accidental overvoltage during test and experimentation.

VOLTAGE-REGULATING POWER SUPPLY WITH CURRENT FOLDBACK

The 2N3055 power transistor has been much used as a series-pass element in linear type voltage-regulating power supplies. Its electrical ruggedness, together with useful voltage, current, and power ratings make it a natural for a wide variety of regulator designs. In order to best exploit the performance potential of this power transistor, due consideration must be given to the error amplifier, the voltage reference, and to a suitable driver stage. A very effective way of providing for these essentials is to utilize a dedicated voltage-regulator IC. These are tailor-made for the purpose; they have self-contained voltage references, and have error amplifiers in the form of high-gain, but stable, differential amplifiers. A transistor driver stage suffices to interface the regulator IC with the high-current 2N3055.

Also, it makes good sense to endow such a voltage-regulated supply with a current-foldover characteristic, such that output voltage and output current "cave in" as response to a heavier than rated or heavier than adjusted for load. Such a feature protects both power supply and load from overload. It is one of the best protective techniques against catastrophic destruction of the pass transistor from a short-circuited output. Not only is it fast acting, but the supply is instantly ready for normal use once the overload condition has been remedied. Thus, the foldback mode is nonlatching. The foldback characteristic is illustrated in Fig. 3-11. Here, 100% loading corresponds to 3.0 amperes, and infinite loading is a short circuit.

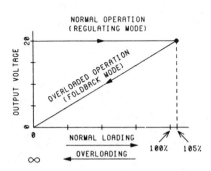

Fig. 3-11. The foldback current-limiting characteristic.

The schematic diagram of such a voltage-regulated power supply is shown in Fig. 3-12. From the above discussion and the block diagram of Fig. 3-13, the functional relationship of its active devices can be readily identified. Basically, the output of an IC voltage regulator is current-boosted by a power transistor used as a series-pass element. Added to this basic configuration is a current foldback circuit which greatly reduces both output voltage and output current in the event of an overload or a short-circuit. This supply is intended to provide 20 volts at up to 3 amperes. However, it is easily modified to produce other voltage-current formats. For example, a 9- to 18-volt adjustable range and a 1-ampere current capability would probably make it more useful for the general needs of solid-state power electronics. As shown in Fig. 3-12, the voltage-adjust potentiometer is intended only for a small adjustment around the nominal 20 volts. (Changing the value of this potentiometer can result in an extended adjustment range for the output voltage, but it would no longer be allowable to draw 3 amperes unless a variac were placed in the primary circuit of transformer T1 in order to reduce power dissipation in the pass transistor.)

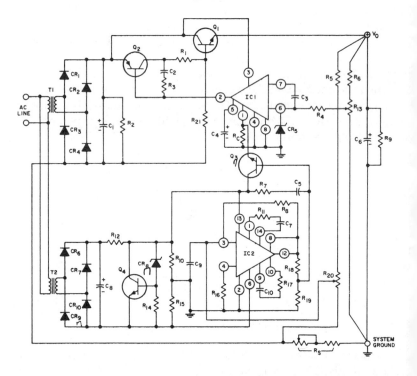

Fig. 3-12. Linear voltage-regulating supply

122 / POWER-CONTROL TECHNIQUES

A word is in order with regard to the drive transistor, Q2. Although Q2 is a PNP type as opposed to the NPN pass transistor, the *combination* of the two form a quasi-Darlington connection. Thus, the current gain of the two transistors working together is the product of their individual current gains. Also, it is coincidental that the series-pass transistor is a 2N3055, and the voltage-regulator IC is a CA3055. The similarity in numerical designation of the two devices should not be allowed to be a point of confusion.

100-WATT 100-kHz SWITCHING SUPPLY

For a number of years, switching regulators have been designed to operate in the vicinity of 20 kHz. At these frequencies, audio-frequency noise from magnetic components was not a problem. At the same time,

T1	Signal Transformer Co., Part No. 24-4 or equivalent	R12	82 ohms, 2 watts, IRC type BWH or equivalent
T2	Signal Transformer Co., Part No. 12.8-0.25 or equivalent	R13	1000 ohms, potentiometer, Clarostat Series U39 or equivalent
CR1-CR4	RCA-1N1614	R16	1200 ohms, 2 watts, wire wound, IRC type BWH or equivalent
CR5	Zener Diode, 1N5225 (3.3 V)	R18	510 ohms, 1/2 watt, carbon, IRC type RC 1/2 or equivalent
CR6, CR7, CR9, CR10	Power Rectifier, RCA-1N3193	R19	10,000 ohms, 1/2 watt, carbon, IRC type RC 1/2 or equivalent
CR8	Zener Diode, 1N5242 (12 V)	R20	300 ohms, potentiometer, Clarostat Series U39 or equivalent
C1	5900 μF, 75 V, Sprague Type 36D592F0758C or equiavlent	R21	510 ohms, 3 watts, wire wound, Ohmite type 200-3 or equivalent
C2	0.005 μF, ceramic disc, Sprague TGD50 or equivalent	RC	240 ohms, 1%, wire wound, IRC type AS-2 or equivalent
C3, C7, C10	50pF, ceramic disc, Sprague 30GA-0.50 or equivalent	RS	(See text for fixed portion); 1 ohm, 25 watts, Ohmite type H or equivalent
C4	2μF, 25 V, electrolytic, Sprague 500D G025BA7 or equivalent	IC1	RCA-CA3055
C5	0.01 mF, ceramic disc, Sprague TG510 or equiavlent	IC2	RCA-CA3030
C6	500 μF, 50 V, Cornell-Dubilier No. BR500-50 or equivlaent	Q1	RCA-2N3055
C8	250 μF, 25 V, Cornell-Dubilier BR 250-25 or equivalent	Q2	RCA-2N5781
C9	0.47 μF, film type, Sprague Type 220P or equivalent	Q3, Q4	RCA-40347
R1	5 ohms, 1 watt, IRC type BWH or equivalent	**Miscellaneous**	
R2	1200 ohms, 1/2 watt, carbon, IRC type RC 1/2 or equivalent	(1 Req'd)	Heat Sink, Delta Division Wakefield Engineering NC-423 or equivalent
R4	100 ohms, 1/2 watt, carbon, IRC Type RC 1/2 or equivalent	(3 Req'd)	Heat Sink, Thermalloy #2207 PR-10 or equivalent
R5	430 ohms, 2 watts, wire wound, IRC Type BWH or equivalent	(1 Req'd)	8-pin socket Cinch #8-ICS or equivalent
R6	9100 ohms, 2 watts, wire wound, IRC Type BWH or equivalent	(1 Req'd)	14-pin DIL socket, T.I., #IC014ST-7528 or equivalent
R7	470 ohms, 1/2 watt, carbon, IRC type RC 1/2 or equivalent	(2 Req'd)	T0-5 socket ELCO #05-3304 or equivalenet
R8	5100 ohms, 1/2 watt, carbon, IRC type RC 1/2 or equivalent		Vector Board #838AWE-1 or equivalent
R9, R14	1000 ohms, 2 watts, wire wound, IRC type BWH or equivalent		Vector Receptacle R644 or equivalent
R10, R15	250 ohms, 2 watts, 1% wire wound, IRC type AS-2 or equivalent		Chassis—As required
R11, R17	1000 ohms, 1/2 watt, carbon, IRC type RC 1/2 or equivalent		Chassis—As required Dow Corning DC340 filled grease

with current foldover. (Courtesy RCA)

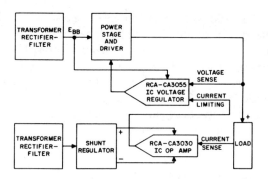

Fig. 3-13. Block diagram of circuit in Fig. 3-12. (Courtesy RCA)

considerable reduction in size and weight was attained over lower-frequency designs. A corollary of this was that the overall power supply also became more cost effective. All this being so, designers longed for the exploitation of higher frequencies. This goal was long impeded by the general lack, or inordinate cost, of devices and components with high-frequency capability. This is no longer the case and high-efficiency regulators operating at 100 kHz and higher can be easily built.

Fig. 3-14 shows the circuit of a 5-volt, 20-ampere switching regulator that operates at 100 kHz and achieves voltage regulation by means of an IC pulse-width modulator. The configuration will be recognized as a forward converter scheme and is noteworthy for its simplicity, considering its respectable performance parameters. Note, in particular, the wide range of AC line voltages that can be used without any need to change input terminals. The half-power to full-power efficiency is about 75% throughout the AC line voltage range of 90 to 260 volts.

Not shown in Fig. 3-14 is a 12-volt power source for operation of the IC pulse-width modulator. A suitable supply for this purpose is shown in Fig. 3-15. It will provide the requisite DC operating voltage over the same line-voltage range that can be accommodated by the switching circuits. Inasmuch as only 50 milliamperes are needed by the IC under worst conditions, almost any transformer with the indicated primary and secondary windings will prove suitable for the purpose. In those applications (probably most) where the exceptionally wide excursions of line voltage do not have to be handled, an even simpler auxiliary power supply can be used; the linear regulator portion (LM117HV and associated components) can be dispensed with and the bridge rectifier can be arranged to supply a nominal 12 to 18 volts of unregulated DC.

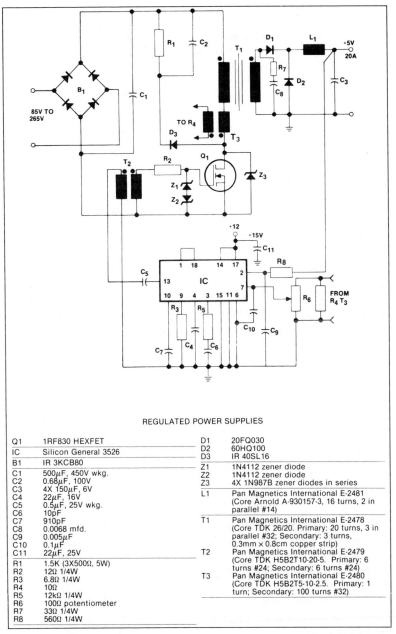

Fig. 3-14. 100-watt switching regulator using a power MOSFET and an IC pulse-width modulator. (Courtesy International Rectifier Corp.)

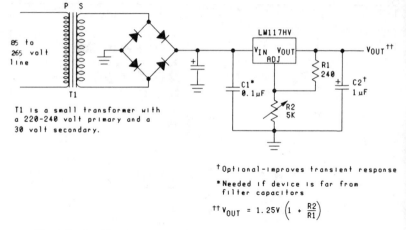

Fig. 3-15. Auxiliary power supply for switching regulator of Fig. 3-14.

In Fig. 3-14, T1 is the main transformer of the basic forward-conversion circuit. It transfers the switched electromagnetic energy from the bridge rectifier to the output rectifier circuit. There is no clamping winding; clamping is provided by D3, R1, and C2. Transformers T2 and T3 interface the MOSFET switching circuit with the IC pulse-width modulator. Such transformer coupling is needed because of the difference in operating voltage levels between the IC and the MOSFET switching converter.

TRACKING A NEGATIVE VOLTAGE

In both design and experimentation it is often awkward to find that a negative voltage in the 10- to 15-volt region is needed in addition to the already present positive supply within that range. Usually, a desirable alternative to building an entirely new negative supply is to reverse the polarity of the already existing positive supply. In essence, this requires a DC transformer. Such a circuit is shown in Fig. 3-16. It not only will furnish the negative voltage, but will track the positive voltage supply. This is desirable inasmuch as most ICs operating from split-polarity power work best with equal-value voltages. Equal-value voltages are beneficial for other loads as well. For example, it is often necessary to preserve DC balance in voltage amplifiers and driver circuits of stereo amplifiers.

In Fig. 3-16 the 555 timer IC is connected as an astable multivibrator and would produce a 50% duty cycle all the time were it not for the feedback path to pin 5 through the NPN transistor. The rectangular wave output at pin 3 alternately turns the P-channel power MOSFET on and off; in so doing, the 1.5-mH inductor in the drain circuit is made to ring, and the negative-going alternations are passed through the diode to the 450-μF filter capacitor. Thus, although the energy derives from the original positive supply, a new DC source is now available which is negative with respect to system ground, or common.

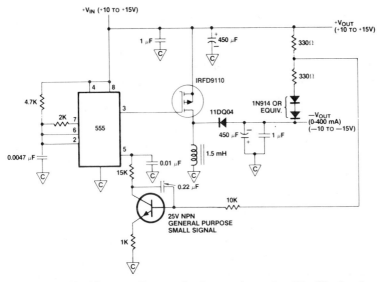

Fig. 3-16. Tracking negative supply. (Courtesy International Rectifier Corp.)

It is the nature of the 555 timer that any imposed change in the voltage already present at pin 5 will alter the timing period. Inasmuch as this IC is connected as a free-running oscillator, this manifests itself as duty-cycle modulation. In this way, the average value of the switching voltage at the power MOSFET gate can be made to follow the positive supply. More specifically, this is brought about via the voltage sensing point at the junction of the 330-ohm resistors, and via the NPN transistor in the feedback path. Note that this transistor needs no external source of collector voltage, this being provided at pin 5 of the 555 IC.

The small power MOSFET can dissipate about 1.5 watts in this application; this calls for some attention to heat removal. The four-terminal, dual in-line package can convey heat from the device into the copper of a printed-circuit board. Two of these terminals are connected to the drain. One should strive for low mounting above the PC board, thick copper board stock, and a mounting area free from nearby heat-producing components. An improvised copper clip to fit over the device can function as a heat sink if some elementary fins are fashioned to transfer heat to the ambient air.

A UNIQUE STEP-DOWN CONVERTER

Early switching power supplies and regulators used individual transistors to accomplish required circuit functions. This led to a high parts count together with reliability and performance problems. Practical implementations often did not come very close to theoretical capabilities. Part of the problem was the difficulty of attaining a circuit layout that abided with good high-frequency practice. Also, with discrete transistors, it was not an easy task to obtain consistent operation in actual production runs. A substantial improvement was realized when op amps became available for the various control functions needed in switching supplies. This led to more compact circuit layouts, to upgrading of reliability and consistency, and to more refined performance. At the same time, it helped bring costs down.

Another quantum jump in switching-supply technology came about with the introduction of pulse-width modulator ICs. With these ICs, all control functions were contained within a dedicated module, and it was only necessary to supply a power switch and a few external passive components. This made possible even higher performance levels; greater efficiencies were attained, together with tighter regulation. Also, because of the compacted circuitry, it became easier to constrain RFI and EMI. Moreover, these dedicated control-modules provided useful pre-engineered sophistications, such as thermal shutdown, dead-time control for push-pull inverters, current limiting, and self-contained voltage references.

Yet another notable advance in switching-supply technology has more recently occurred. This involves ICs comprising pulse-width modulator circuitry and the power-output stage on a single monolithic chip. Previously, the closest approach to this goal was achieved via hybrid technology. The Lambda LAS 6380 switching regulator series is the pioneer device making use of the monolithic format. The LAS 6380 not only contains an 8-ampere Darlington output transistor, but its control circuitry contains a temperature-stabilized voltage reference, internal current-limit protection, internal thermal shutdown, a DC-to-200-kHz oscillator, an inhibit/enable control pin, and double pulse suppression logic.

A step-down converter circuit using the LAS 6380 is shown in Fig. 3-17. This unique regulating power supply is intended to convert a nominal 20-volt source to 5 volts at up to 8 amperes. The switching rate is 40 kHz, so inductors and capacitors are of reasonable size. Because of this high switching rate, free-wheeling diode D1 is specified as a Schottky diode. For the most effective utilization of the capabilities of D1, its cathode should be as close as possible to pin 8 of the LAS 6380; its anode should be as close as possible to the ground point of the 0.1-μF capacitor in the pin 4 circuit. Ground-loop avoidance will be approached by placing C_{IN} close to pin 1. Maximized thickness should be used for printed-circuit copper runs carrying high currents.

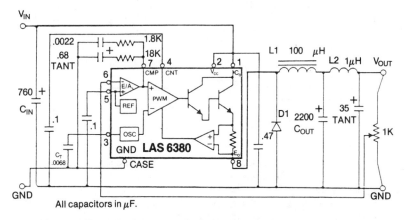

Fig. 3-17. Step-down converter utilizing combined PWM control and power switch. (Courtesy Lambda Semiconductors)

The general performance characteristics of this circuit are depicted in Table 3-1.

A MULTIPLE-OUTPUT SWITCHER

A particularly useful power supply is one with +5-volt and ±12-volt outputs. This happens to be a voltage combination capable of satisfying the requirements of many TTL logic systems, for such systems often include linear ICs which operate from ±12-volt sources. The conventional multioutput switching supply usually contains a separate choke and output transformer. The novel circuit shown in Fig. 3-18 dispenses with one of these core components. It will be seen that the choke and output transformer are combined as a single unit. Of the three output voltages, the +5 output is the most tightly regulated. This makes sense, for logic circuit operating voltage is much more critical than is DC operating voltage for linear ICs. The +5-volt output has an output-current capability of 5 amperes; the two 12-volt outputs can each supply 125 milliamperes—more than enough for the usual requirement for linear ICs.

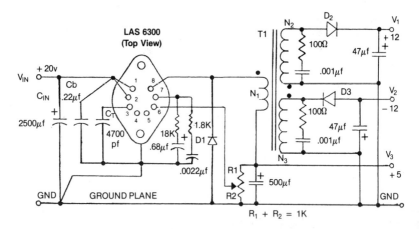

Fig. 3-18. Triple-output switching supply.
(Courtesy Lambda Semiconductors)

The LAS 6300 is a sophisticated subsystem, as can be appreciated from the block diagram of Fig. 3-19. It contains its own internal linear regulator and voltage-reference source. There are also provisions for either internal or external current limiting, thermal shutdown, and remote control. Pulse-width modulation is accomplished in the comparator which samples signals from the error amplifier and the oscillator. Interestingly, current limiting is brought about by shifting the frequency of the oscillator.

This design approach takes the headaches out of switching-supply implementation inasmuch as the parts count is minimal, and no external power transistor is needed. A suitable heat sink is the model 5759B-15 or the Model 5006B-15 made by AAVID Engineering, Inc. of Laconia, New Hampshire. If due consideration is given to the phasing of the windings on T1, and Schottky diodes are used for D1, D2, and D3, no difficulties should be experienced in obtaining proper operation. (This is not an easy statement to make for switching supplies comprising a large number of transistors and discrete parts.)

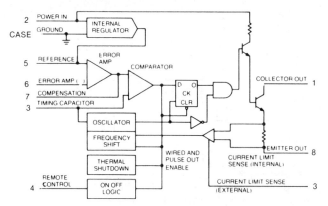

Fig. 3-19. Block diagram of LAS 6300 switching regulator.
(Courtesy Lambda Semiconductors)

MULTIPLE-OUTPUT, 230-WATT, 50-kHz REGULATED SUPPLY

Switching power supplies, especially those intended for providing the requisite DC voltages for computers, generally don't qualify as simple weekend projects for the hobbyist. Yet on a comparative basis, the state-of-the-art 230-watt, 50-kHz regulated supply shown in Fig. 3-20 is considerably less involved than most traditional designs. The block diagram of Fig. 3-21 will help drive home the essentially straightforward functions of the several function sections of the supply. Referring again to the schematic diagram of Fig. 3-20, it should be noted that the entire bottom half of this circuit comprises the "brains," or switching logic of the supply. Further simplification can be realized by observing that the switching-logic function takes place within a dedicated pulse-width modulation IC, the Signetics NE5560. The associated blocks merely perform interface and auxiliary functions for the NE5560 pulse-width modulator.

The duty-cycle modulated buck regulator consists of the two IRF720 power MOSFETs. (These devices are called HEXFETs by International Rectifier Corp.) This regulator circuit can operate efficiently at any duty cycle between zero and almost one-hundred percent. The rectified output from the power switch then controls the 50-kHz output level of the bridge-configured power switch comprised of four IRF720 HEX-

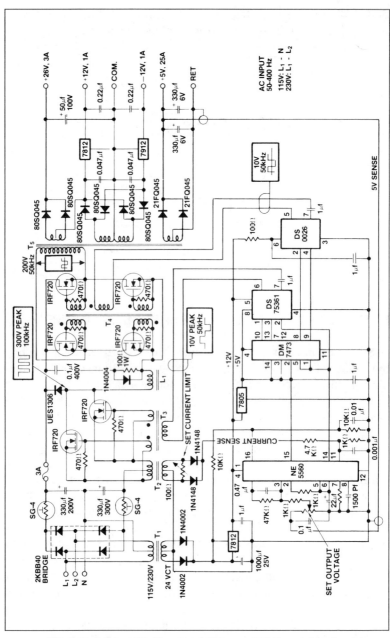

Fig. 3-20. Schematic diagram of 230-watt, 50-kHz regulated power supply. (Courtesy International Rectifier Corp.)

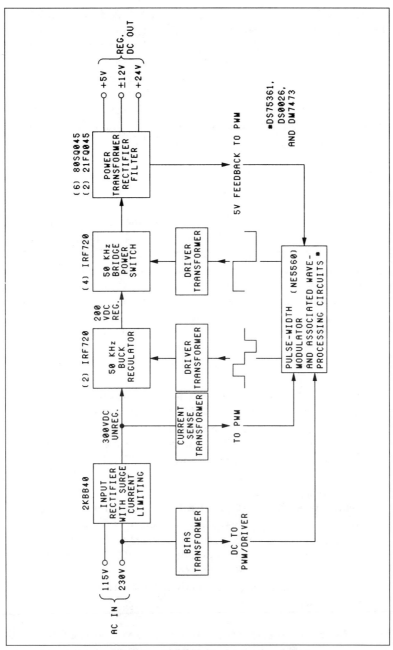

Fig. 3-21. Block diagram of circuit in Fig. 3-20. (Courtesy International Rectifier Corp.)

FETs. This output stage operates at a constant 50% duty-cycle, and delivers a 50-kHz square wave to the rectifier system.

Although bipolar transistor supplies have been successfully designed for 50 kHz and and higher switching rates, the requisite transistors border on the exotic and considerable attention must be given to snubbing circuits and other protective techniques. The HEXFETs are intrinsically capable of almost negligible switching loss at much higher rates than 50 kHz. Additionally, they are immune to thermal runaway, are easy to drive, and can be directly paralleled (no ballast resistors) for greater output power. Usually nothing needs be done to the drive circuitry as the result of paralleling.

The salient features of the 50-kHz HEXFET switching supply are:

- There are no 60-Hz magnetics. Off-the-line operation can accommodate 115- and 230-volt lines over a frequency range of 50 to 400 Hz. For 115-volt operation, the power line is connected across terminal N and either terminal L1 or terminal L2. The input rectifier system then operates as a voltage doubler, producing a nominal 300 volts of unregulated DC. For 230-volt lines, the AC power is connected to terminals L1 and L2. The input rectifier system then operates as a conventional bridge circuit, again developing approximately 300 volts DC. In either instance, the SG-4 thermistors provide soft-start characteristics for the supply.
- The output rectifier-system uses no filter chokes. This is unnecessary because the 50-kHz square wave always maintains its 50% duty cycle. Moreover, the rise and fall times of the square wave are extremely fast. Capacitor filtering suffices for the 5- and 26-volt outputs. The plus and minus 12-volt outputs benefit further from the electronic filtering produced in the 7812 and 7912 linear regulators.
- Schottky rectifier diodes are used. These have low switching losses, as well as low forward voltage drops. All rectifying circuits are full wave. Note that full-wave rectification of a perfect square wave would produce pure DC with no filtering. In this supply, the HEXFETs and the Schottky diodes develop a very good square wave and filtering demands are relaxed. The 50-kHz switching rate also helps in this regard.

The feedback loop from the output is derived from the 5-volt rectifier circuit. However, through the mutual coupling of the windings on

T5, regulation is imparted to the other output voltage. Inasmuch as the plus and minus 12-volt outputs contain their individual linear regulators, this is of no consequence for these voltages. The 26-volt output is, however, one that is slaved to the regulation of the 5-volt output. Actually, the regulation of the 26-volt output will not be quite as good as that of the 5-volt output. Inasmuch as the 5-volt supply to logic circuits tends to be critical in some instances, it is more important to design for maximum stabilization of the 5-volt output than the others. In any event, it is generally found that 26-volt sources are less demanding of tight regulation than either the 5- or 12-volt sources in the operating of computer and logic circuits.

A BIMOS BRIDGE INVERTER

The inverter to be described operates from 750 volts DC and delivers 10 amperes to a load. The switching rate is 25 kHz. As pointed out in Chapter 1, such a combination of performance parameters is useful in industrial applications. However, thyristors, bipolar transistors, including Darlingtons and power MOSFETs, all exhibit profound shortcomings for service in this domain, these being cost, reliability, efficiency, or speed. By a fortuitous coincidence, a combination of two commonplace devices working together in a cascode circuit—the BIMOS power switch—yields performance ordinarily attainable via the use of exotic and costly devices.

The 25-kHz switching portion of the inverter is comprised of four such BIMOS power switches in a bridge configuration. This is shown in Fig. 3-22. It can be seen that the bipolar transistor is operated as a common-base switch, and that its emitter current can be interrupted by the MOSFET. Bipolar transistors have much better frequency and SOA ratings in the common-base circuit than in the more conventional common-emitter circuit. Because of the presence of the MOSFET, the overall power switch is easy to drive. (Conversely, a common-base power switch, by itself, would be very difficult to drive.) Note the symmetry of the bridge—all four BIMOS circuits are identical.

These BIMOS switches are a little different from the simplified circuit discussed in Chapter 1. The difference pertains to the method of obtaining turn-on bias for the bipolar transistor. Initially, this bias derives from the charge stored in capacitors C1, C3, C5, or C7. However, once collector current flows, the remainder of forward base bias is obtained from the collector-current transformers, CT1, CT2, CT3, or

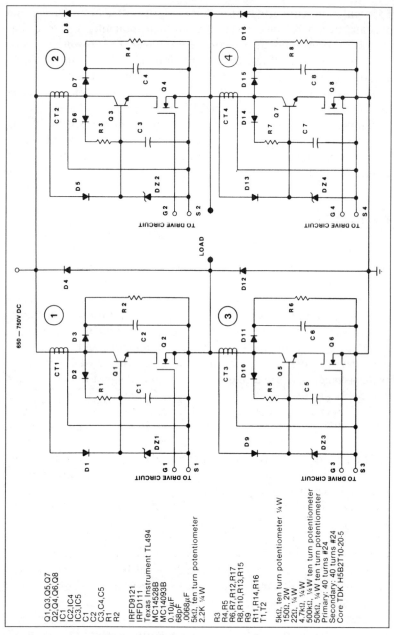

Fig. 3-22. The BIMOS bridge circuit. (Courtesy International Rectifier Corp.)

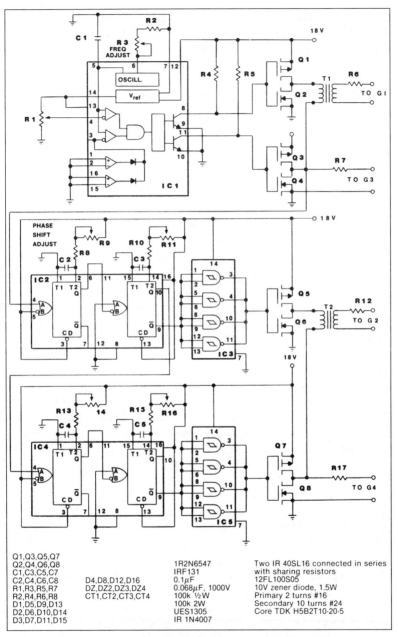

Fig. 3-23. Control and drive circuit for the BIMOS bridge.
(Courtesy International Rectifier Corp.)

REGULATED POWER SUPPLIES / 137

CT4. The output of the bridge converter is a nominally 25-kHz AC wave which can be controlled both in frequency and in duty-cycle by the control circuit. The BIMOS bridge works well with resistive and inductive loads. (A 4- or 8-μF capacitor may be placed in series with some inductive loads to prevent magnetic saturation from DC components accompanying slight waveform imbalance.)

The control and drive circuit for the bridge is shown in Fig. 3-23. At first glance, this circuitry appears to be quite complicated. This, however is because the draftsman detailed the logic devices comprising the ICs. For the practical purposes of construction and testing, these ICs can be considered as "black boxes." Such a block approach greatly simplifies the appearance of the circuit and is justified on the premise that we do not have access to the internal circuits of these ICs. On the other hand, the draftsman's diligence is rewarding, for it is instructional to see what is basically inside of the ICs.

IC1 is a regulator control circuit. It renders unnecessary the oscillator, voltage reference, pulse-width modulator, output stages, and other electronics commonly needed in switching-type power supplies. IC1 is the source of the two 180-degree phase-displaced waves needed for bridge gates 1 and 3. IC1 also supplies a small deadtime in these waves to protect against simultaneous conduction of the power switches. The waves needed for bridge gates 2 and 4 are then derived from these basic waveforms.

IC2 and IC4 are retriggerable, resettable monostable multivibrators. Their functions are to provide adjustable delays (phase shifts) on the edges of the waveforms for bridge gates 2 and 4. By this means, the duty cycle of the voltage waveform delivered to the load can be varied manually. Controlling the duty cycle of an AC wave in this fashion is an efficient way to control its RMS amplitude.

IC2, IC5, and Q1 through Q7 are drivers for the BIMOS power switches in the bridge circuit. IC3 and IC5 appear to add complexity to the operation of the circuit, but this is deceptive. These ICs are quad, two-input NAND Schmitt triggers, but they are not deployed in such a way as to exploit their logic potential. It will be noted that the input leads are all connected in parallel, as are all of the outputs. Accordingly, the simple function of these ICs is to act as a low-impedance source of the fast rise and fall waveform for ultimately driving two of the BIMOS power switches. Drive of the BIMOS switches takes place through pairs of MOSFETs, which must be inputted from a low-impedance source in order to preserve the steep edges of the waveform.

This inverter is easy to troubleshoot inasmuch as the control and drive circuit may readily be tested separate from the bridge circuit. Assuming all ICs and active elements are properly connected and in good working order, a possible source of malfunction could be incorrect phasing of drive transformers T1 and T2.

Table 3-1. Performance Characteristics of the Step-down Converter of Fig. 3-17. (Courtesy Lambda Semiconductors)

PARAMETER	CONDITIONS	LIMITS MIN	TYP	MAX	UNITS
Output Voltage Tolerance V_{OUT} Trimmed to 5 Volts	$T_J = 0°C - 125°C$ $V_{IN} = 12 - 30V$ $I_{OUT} = .25 - 8A$		±2	±3	%
Output Voltage Initial Accuracy	$I_{OUT} = 1.0A$		±3	±5	%
System Conversion Efficiency	$I_{OUT} = 4.0A$ $I_{OUT} = 8.0A$	73 70	77 74		% %
Output Noise and Ripple	$V_{IN} = 30V$		.03 .012		Volts Peak Volts RMS
	V_{IN} 25V + $5V_{PK}$ @ 120 HZ		.03		Volts Peak
Output Current Limit			10.5		Amps
Output Dynamic Response Turn-On Overshoot Peak Duration	$I_{OUT} = .05A$		500 40		mV mS
Peak Duration	$I_{OUT} = 8A$		0 0		mV mS
Unit-Step Load Change Peak Duration	I_{OUT} from short circuit to 8A		0 0		mV mS
Peak Duration	I_{OUT} from 8A to .05A		500 60		mV mS

$V_{CC} = V_{IN} = 20$ VDC, $V_{OUT} = 5V$, $I_{OUT} = 8A$, Fsx = 40 kHZ $T_A = 25°C$, unless otherwise stated. See Test Circuit

CHAPTER 4

Control of Electric Motors

Inherently intriguing is the control of electric motors via the brains and nerves of electronic devices and circuits. In such methodology, the "muscle," represented by the motor, obediently follows the operating instructions conveyed by servo, feedback, or digital-programming techniques. Many basic ideas underlying such control date back to an earlier era. However, electron tubes just didn't have the right combination of electrical and physical characteristics to qualify for widespread implementation in motor-control systems. With the advent of the first power transistors, interest was revived in this field, for we then had an "electronic rheostat" with the kind of parameters suitable for motor control. Much was accomplished despite the limitations and inadequacies of the early transistors. Obviously, the wheels of progress had to turn for a number of years until the capability and reliability of power transistors finally merited recognition as a really practical motor-control element. While such progress was in effect, the SCR made its debut and quickly caught on as a valuable adjunct in motor-control systems. Today, these semiconductor devices and their derivatives provide an extensive family of motor-control elements with previously unthinkable ratings in voltage, current, power, and switching speed.

The motor-control applications discussed in this chapter deal with both DC and AC motors. The circuits tend to be flexible in that the scaling techniques needed for larger, smaller, or somewhat different motors than those designated are quite straightforward. This is very much because of the wide availability of semiconductor power devices with an extensive range of parameter ratings. It also follows from the accommodating nature of many control techniques. For example, power MOSFETs often can be paralleled with no need to provide increased drive power. And often a larger SCR or triac can be substituted for the stipulated one with retention of the original trigger circuit.

Motor control is an exceedingly important field; it bears relevantly on innumerable industrial processes, on traction vehicles, on automotive and aircraft auxiliary functions, and on the developing technology of robotics. Then, too, there are many applications involving toys and various hobbies.

Last, but not least, is the interesting phenomenon of electric control of motors with regard to motor behavior. The classic characteristics pertaining to the various types of electric motors are often drastically modified; thus, the textbook-described relationships linking speed, torque, power, and starting performance may be deliberately or inadvertently altered in control circuits using various sensing and feedback techniques. In general, this expands the flexibility and reduces the limitation of motors.

A VARIABLE-SPEED DC MOTOR CONTROLLER

The DC motor has long been known for the ease with which speed control can be implemented. All that needs be done is to vary the voltage applied to the armature (assuming a permanent-magnet type). Traditionally, this has been accomplished by means of a rheostat. The shortcoming of this method is that the overall system is very inefficient by virtue of the power dissipation in the rheostat. Note that the same drawback prevails when the variable-resistance rheostat is replaced by a solid-state rheostat, such as a transistor. A more efficient approach is to chop the DC and obtain the variation in applied voltage by varying the duration of the ON time of the chopper. The principle of operation for doing this with power transistors is old hat, for it is merely an adaptation of the pulse-width modulation concept long used in switching power supplies. However, this approach has often suffered from inordinate circuit complications.

A relatively simple PWM speed control for DC motors is shown in Fig. 4-1. This circuit is simpler than might be assumed from first glance, for the four op amps are all part of a single IC module. All that is needed is the IC, the power MOSFET, and a few passive components. Previous implementations of this control concept usually required a tachometer. None is needed in this circuit, however.

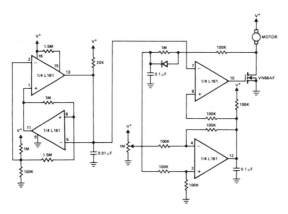

Fig. 4-1. Efficient circuit for controlling the speed of a DC motor.
(Courtesy Siliconix Inc.)

Speed sampling is accomplished by monitoring the counter emf of the motor. This sensing is done by the lower right-hand op amp through the 100K resistor connected to the motor armature. (The sensing takes place when the power MOSFET is turned off.) The upper right-hand op amp is the PWM modulator. It delivers a rectangular wave to the motor, with a duty cycle determined by the DC voltage level applied to its noninverting terminal by the previously described sense amplifier, the lower right-hand op amp. The inverting input terminal of the PWM receives a triangular wave from an oscillator comprised of the two left-hand op amps. Thus, the circuit is straightforward—the PWM is an op amp fed by a variable DC voltage and a triangular wave.

Speed control occurs because of the change in the average value of armature current that can be brought about by adjusting the one-megohm potentiometer associated with the sensing amplifier. Unlike some rheostatic speed controls, this circuit maintains the motor speed constant for any adjusted speed. Even though there are variations in applied motor voltage, in mechanical loading of the motor, or in motor temperature, this speed-control circuit will maintain the set speed at a constant value.

EFFICIENT SPEED CONTROL OF PERMANENT-MAGNET DC MOTORS

The motor-control circuit depicted in Fig. 4-2 brings together three noteworthy devices that have been instrumental in simplifying, ruggedizing, and improving electronics control systems. These are the permanent-magnet DC motor, the power MOSFET, and CMOS logic ICs. The permanent-magnet motor, utilizing modern ceramic magnetic material has enabled the manufacture of small compact motors with much higher starting torque than was previously achieved in motors of this type. Other attributes of the motor have also been upgraded, such as speed constancy under varying load, and commutation. These motors have only two terminals; there is no wound field. To reverse rotation, it is only necessary to transpose the power lead connections to the brushes.

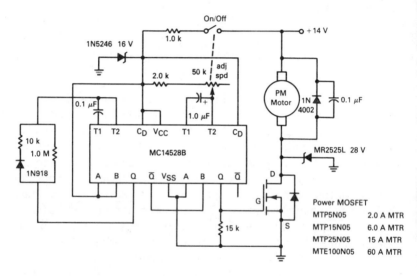

Fig. 4-2. Efficient speed control of permanent-magnet motors.
(Courtesy Motorola Semiconductor Products, Inc.)

The power MOSFET is a very nice device for power control. Unlike the bipolar power transistor, it is immune to secondary-breakdown destruction, and does not self-destruct from thermal runaway. (That is not to say that similar catastrophic destruction could not be brought about by gross overloading and extremely abusive operation. Note, however, that the temperature coefficient of resistance of these devices is opposite that of bipolar transistors; this implies that the power MOSFET trys to *limit* excessive power dissipation.) And, whereas large bipolar power transistors are difficult to drive, the input impedance of the power MOSFET at DC is, for practical purposes, infinite. Finally, the switching performance of these devices is superb—it is generally easier to get efficient high-speed operation with a power MOSFET than with bipolar power transistors. Although price and relatively high ON resistance frequently loomed up as obstacles, these barriers have been largely overcome and continued developmental progress is the order of the day.

CMOS logic circuitry and linear ICs enable sophisticated circuit elements to be integrated and marketed at low cost. It is often no longer profitable, or electrically advantageous, to laboriously construct circuit functions and subsystems with numerous discrete elements and their accompanying high-parts count in passive components.

In making use of these compelling features, the circuit of Fig. 4-2 also provides considerable design flexibility inasmuch as power MOSFETs of appropriate power capability can be selected for use with different size motors. (It is about as easy to drive a large power MOSFET as a small one.) Moreover, power MOSFETs can be readily paralleled without the power-wasting ballast resistors needed when paralleling bipolar power transistors. This, too, is a consequence of their temperature coefficient—individual devices in a parallel-connected group do not try to hog the current.

The MC14528B is a dual monostable multivibrator. In the circuit of Fig. 4-2, one half of this IC is connected to function in the astable mode, thereby becoming a pulse generator. The other half is deployed as a monostable multivibrator with adjustable pulse width. This adjustment varies the average current allotted to the motor, and therefore provides manual control of speed. Inasmuch as the power MOSFET is either ON or OFF, there is relatively little power lost, and overall efficiency is high, even at slow speeds. During inter-pulse time, the motor relinquishes energy stored in its magnetic field via the flow of current through the free-wheeling diode connected across its armature terminals. In so doing, the motor experiences more continuous torque than would otherwise be the case, and operation is smoother and more efficient because of it. However, be certain that the current capability of this diode is compatible with the size of the motor.

The combination of CMOS control logic and the power MOSFET results in very low standby power drain, a welcome feature in battery-operated systems. In most instances, it will be found that EMI and RFI is much less than for thyristor-type motor controllers.

CONTROL OF DC MOTOR SPEED WITH PWM MODULE

Many fractional-horsepower DC motors require from 5 to 35 volts and consume up to 5 amperes. This range falls within the capability of the Lambda LAS 6300 pulse-width modulator. Because this IC has this power capability, and because a sufficiently high frequency can be used so that the inductance of the motor armature itself serves to integrate the current through it, an extremely simple and efficient motor-speed control can be devised. No external power transistor or chokes are needed, and only a few passive components are required. The circuit of such a motor-speed control is shown in Fig. 4-3. No feedback is used; a DC voltage applied at V_C controls the speed. This voltage may be derived from a potentiometer connected across the DC voltage source, V_M. The switching rate of 25 kHz is set by the 0.025-μF capacitor connected to pin 3.

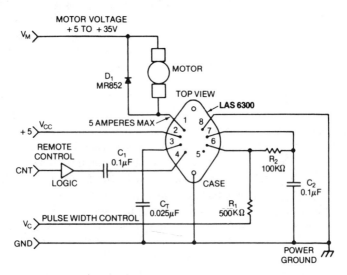

Fig. 4-3. Speed control of fractional horsepower DC motor with a PW IC module. (Courtesy Lambda Semiconductors)

The functional block diagram of the LAS 6300 is shown in Fig. 4-4. It provides insight into the nature of the speed control, as well as some of the self-protection provisions of this IC. Pin 4 provides a useful control function. Note that in the motor-control circuit of Fig. 4-3, pin 4 is connected to a coupling capacitor and to a logic device. This setup is used in the following manner: A momentary logic 0 turns the LAS 6300 on, whereas a momentary logic 1 turns it off. Thus, the motor may be remotely turned on and off.

An interesting aspect of this motor-speed control is that excessive motor current causes the LAS 6300 to go into its current-limiting mode.

This operational mode confers protection to both the IC and the motor by lowering the oscillator frequency, thereby reducing the duty cycle of the motor current. On the other hand, thermal shutdown also occurs automatically, but in this protective measure the output current pulses are totally inhibited. The experimenter should be aware, however, that thermal shutdown will not occur if the DC resistance from pin 4 to ground is less than 5000 ohms.

This is a sure-fire control circuit, but be careful with selection of free-wheeling diode D1. It must be either a Schottky diode or a fast-recovery junction diode—rectifier, or garden-variety diodes will produce inefficient motor operation and will endanger the power-transistor stage in the LAS 6300. The best DC motor for speed control is the modern permanent-magnet type using ferrite or ceramic field

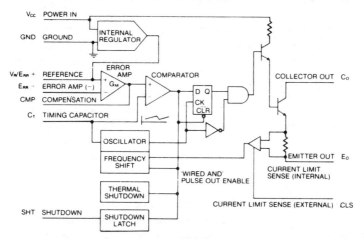

(A) Block diagram.

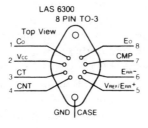

(B) Top view pinout.

Fig. 4-4. Functional block diagram and pinout of LAS 6300 PWM controller. (Courtesy Lambda Semiconductors)

magnets. The experimenter will also find series motors suitable for some applications. With the PM motor, reversal of rotation is conveniently achieved by reversing the connections to the motor. With a series motor, either (but not both) field or armature connections must be reversed. A motor with a wound shunt field can also be used; the shunt field can be connected across the armature terminals. Conversely, a separate DC source for the shunt field can be provided. In any event, reversal of rotation can be brought about by simply transposing the shunt-field connections. Shunt-field current is very low relative to armature current, so this type of DC motor is easy to implement.

UNIQUE BIDIRECTIONAL SERVO DRIVE SYSTEM

A servo drive system involving rotational motion needs power gain, bidirectional motion, and a dead band; the latter is generally required to prevent oscillation, overshoot, or other instabilities. The conventional design approaches use linear amplifiers and special motors, such as two-phase induction types. A simpler and more economical scheme is shown in Fig. 4-5. Here, the behavior of such a servo drive system is simulated, but with a DC series motor or universal motor, and without amplifiers in the conventional sense. In order to gain insight into the operation of this circuit, it will prove expedient to first consider some clever uses of devices and circuitry elements not commonly encountered in power control.

In Fig. 4-6A a DC series motor is shown connected to a polarity-reversing circuit. The direction of rotation *cannot* be changed by reversing polarity. In order to reverse rotation, the connections to either, but *not* both, the armature or the field winding would have to be transposed. However, by surrounding the motor with steering diodes as depicted in Fig. 4-6B, such transposition is not necessary; in this case, reversal of applied polarity, as provided by the dpdt switch, suffices to reverse rotation. It is also interesting to note that if the battery or DC source were replaced by an AC source, the motor in Fig. 4-6A would run, whereas the motor in Fig. 4-6B would not run—its average torque would be zero. Thus, by appropriate use of the steering diodes, some of the natural characteristics of the series motor can be electronically modified.

Another unusual deployment of components is illustrated in Fig. 4-7 which shows the triggering characteristics of diac-triac combinations under usual and unusual conditions. In Fig. 4-7A, the waveforms of a

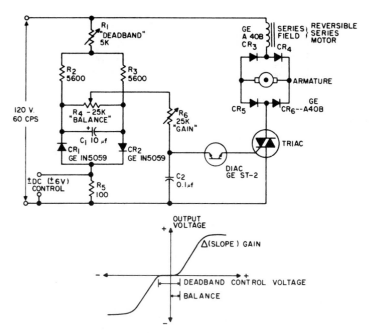

Fig. 4-5. Bidirectional servo drive system. (Courtesy General Electric Company)

diac and a triac in conventional phase-control circuits, such as light dimmers, are shown. The salient feature is the AC symmetry of these waveforms. A phase angle of approximately 90 degrees is represented in the drawings; but no matter what the phase angle, the triac waveform remains symmetrical, i.e., it is always a pure AC wave with *no* DC component. In the situation depicted in Fig. 4-7B, a DC bias is applied to the diac so that voltage breakdown occurs for only one polarity of the AC wave. As a consequence, the triggered waveform of the triac is assymetrical and has an obvious DC component. If the DC bias applied to the diac is reversed in polarity, the situation is then as shown in Fig. 4-7C. Thus, a low-power DC signal can control the DC polarity of a relatively high-powered DC output from the triac. In a sense, the triac behaves somewhat like an SCR when used in this manner in that SCR action provides rectification, but conventional use of the triac does not.

By combining these unique implementations of the motor and triac, the operation of the circuit of Fig. 4-5 becomes straightforward. Essentially, it can be viewed as combining an electronically modified motor with a polarity-selectable high-gain DC source. Although it is not commonplace to think of a thyristor, such as a triac, as providing power am-

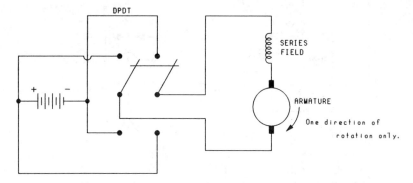

(A) Reversing polarity to series motor does not reverse direction of rotation.

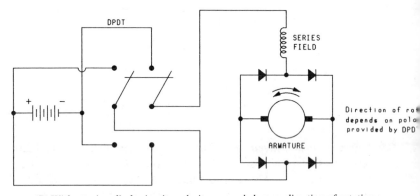

(B) With steering-diode circuit, polarity reversal changes direction of rotation.

Fig. 4-6. How the direction of a series motor can be made polarity responsive.

plification, it is only because such a viewpoint is not particularly convenient in most applications. In this particular case, the concept of power amplification is meaningful inasmuch as a low-level variable-polarity DC signal controls the activity of a powerful motor in the same fashion as is ordinarily done with a linear or analog amplifier.

The allusion to a reversible series motor in Fig. 4-6 implies that this is the optimum type of motor for the purpose; actually, *all* DC series motors, as well as all universal motors, will work in this setup. If, however, a series motor is not designated as reversible, it often happens that brush alignment is off geometric center in order to favor commutation for one direction. If such a motor has adjustable brushes, it is generally best to reposition them. If not, it may be found that performance is satisfactory nevertheless.

In most servo systems using series motors, it has been necessary to disconnect the internal connections between field windings and the armature. This surgery is not always trivial, especially in universal motors where the manufacturer assumed these connections would be permanent. Obviously, a compelling feature of reversible series-motors is that such motors can be used in their natural state. Incidentally, there appears to be little reason why the competent experimenter couldn't adapt features of this scheme for other types of DC motors. With a permanent-magnet motor, the steering diodes would not be needed. With a shunt-field motor, it probably would be necessary to insert a rudimentary filter between the triac and the field winding.

This servo driver can readily be converted into a full-fledged servo system by the addition of an appropriate negative-feedback loop and a reference voltage. A straightforward implementation would involve a small DC tachometer mechanically coupled to the motor, and a resistive network for combining the tachometer, reference, and control voltages at the input of the system.

FULL-WAVE PHASE CONTROL OF MOTOR WITH AN SCR

An interesting phase-control circuit for universal and DC motors is shown in Fig. 4-8. Here, full-wave power is applied to the motor even though only a single SCR is used. The basic principle underlying this implementation can be understood from a contemplation of the simplified circuit of Fig. 4-9. This circuit shows that a bridge rectifier appropriately associated with the single SCR enables the overall circuit to apply phase-controlled AC to the load. The waveforms will be seen to be essentially identical to those ordinarily provided by antiparallel connected SCRs, as well as to the more familiar triacs. Possible justifications for utilizing this basic arrangement rather than a triac could entail one or more of the following factors:

- Both SCRs and rectifying diodes are capable of greater electrical ruggedness than the triac. Both of these solid-state devices can be readily obtained in higher voltage, current, and power ratings than triacs.
- Both SCRs and rectifying diodes can be obtained with greater frequency capability than triacs.
- Comparing cost on a dollar per watt, or dollar per peak ampere basis, some applications would merit consideration of the SCR plus rectifier bridge approach.

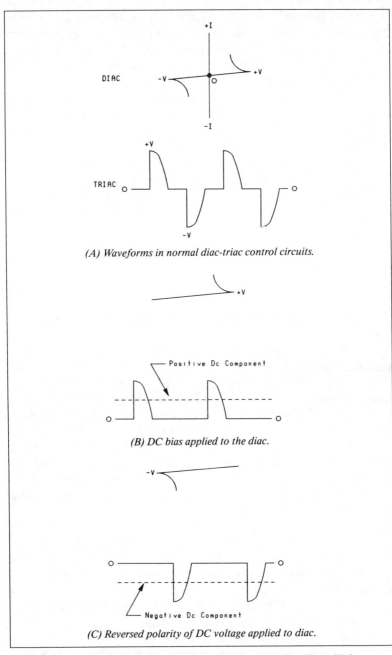

(A) Waveforms in normal diac-triac control circuits.

(B) DC bias applied to the diac.

(C) Reversed polarity of DC voltage applied to diac.

Fig. 4-7. Idealized diac and triac waveforms as function of DC bias on the diac.

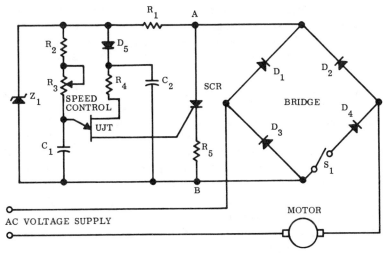

SCR Motorola MCR 808-4		Z1	1N751A—5.1V, 400mW Zener Diode
		R1	18K, 2W
UJT MU970		R2	3.9K, 1/2W
		R3	50K, Potentiometer
D1,D2	1N3493—200V, 18A Rectifier	R4	330Ω 1/2W
D3,D4	1N3493R—200V, 18A Reverse Polarity Rectifier	R5	See table
		C1	.1μfd
D5	1N4001—50V, 1A Rectifier	C2	10μfd, 10V

	NOMINAL R5 VALUES		
MOTOR RATING	R5		
(AMPS)	(OHMS)	(WATTS)	R5 = 2 / IM
2	1	5	
3	0.67	10	
6.5	0.32	15	IM = Max Rated Motor Current (RMS)

Fig. 4-8. Motor-control circuit featuring speed regulation and full-wave operation from a single SCR. (Courtesy Motorola Semiconductor Products, Inc.)

The objective of providing full-wave rather than half-wave power to the motor is to obtain smoother performance, wider control range, and greater maximum torque or speed. Also, if speed regulation is involved, the full-wave circuit generally provides better stabilization. The availability of bridge-rectifier modules enhances the convenience of constructing such a circuit.

In the circuit of Fig. 4-8, the firing of the unijunction transistor is advanced more as increased motor current flows through resistor R_5. Thus, when the motor is loaded and tends to lose speed, it generates less counter emf and the manifestation of this is greater motor current. Greater motor current develops a higher voltage drop across R_5; the polarity of this voltage drop is such that the firing of the unijunction transistor is attained sooner than the time that was initially set in by adjustment of R_3. In other words, as the motor experiences more mechanical loading, it takes less time for capacitor C_1 to charge to the firing potential of the unijunction transistor. The result of this timing advance in the UJT relaxation oscillator is earlier per cycle triggering of the SCR, and therefore increased motor voltage. This, in turn, is accompanied by increased motor torque, which is evidenced by higher speed. Thus, the behavior of the system is such that a nearly constant speed characteristic is maintained with respect to motor loading; the corrective tendency works both ways—a relaxation in motor loading will also invoke action preventing any appreciable increase in speed.

Explanations are in order regarding some of the details of this circuit. Zener diode, Z_1, in conjunction with R_1, forms a DC source for the operation of the UJT relaxation oscillator. However, after the triggering of the SCR, the available voltage at points A-B is less than the breakdown voltage of the zener diode. This would tend to make the UJT vulnerable to undesirable firing—lowering of the interbase voltage in these devices reduces the required firing potential. To safeguard against this, a memory circuit is formed of diode D_5 and capacitor C_2. Here, C_2 retains the approximate voltage developed across Z_1 prior to the drop of

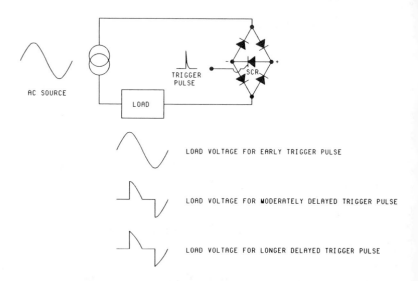

Fig. 4-9. Simple circuit for full-wave phase control of load power using one SCR.

voltage AB. Inasmuch as C_2 is a large capacitor, it takes over as DC source for the interbase circuit of the UJT during conduction of the SCR.

Nothing has been said about commutation of the SCR. Actually, no special provision is needed; conduction ceases in the SCR at the zero crossing of the AC voltage wave. That is why there is no filter capacitor across points A and B—such filtering would produce good direct current for the motor, but the circuit would hang up and be inoperative. In order for the circuit to work, the SCR must automatically turn off at the end of each half cycle. Fortunately, series and universal motors can be reasonably happy on AC waveforms.

Switch S_1, in the rectifier-bridge arm, provides the option of half-wave operation. Under most circumstances, superior motor operation is obtained from the full-wave format. However, under some load conditions, it may happen that instabilities such as hunting, or erratic speed changes, may be experienced; changing to half-wave operation can sometimes beneficially alter the feedback parameters and alleviate such malperformance. Another use for half-wave operation is with shunt, or permanent-magnet DC motors. Such motors will operate with their armatures impressed with pulsating DC or unidirectional current, but *not* with AC waveforms.

It is not necessary to duplicate the designated semiconductor devices of this circuit; the components, with the exception of R_5, are not critical. Much will depend upon the type of motor used and the load variations it is subjected to. Although the table associated with the parts list provides nominal values for R_5, it would be a good idea to use an adjustable wirewound type of resistance, or a rheostat. In any event, this component should be of good quality and conservative rating. Changes in resistance because of temperature coefficient or faulty contacts are likely to adversely affect speed regulation.

CONTROLLING AC AND DC MOTORS WITH A POWER OP AMP

The RCA HC2500 is a hybrid operational amplifier with considerable power capability; it can provide 100 watts of RMS power to a load, and can deliver 7 amperes of peak current. The total DC supply voltage can be as high as 75 volts and either single or split DC supplies can be accommodated. Such ratings make this device useful for control of small motors. Implementation is simple inasmuch as no external power transistors are needed, as would be the case with conventional op amps.

Two experimental motor-control applications are shown in Fig. 4-10. The setup depicted in Fig. 4-10A allows the speed of an induction motor or synchronous motor to be controlled via frequency variation. The idea is to feed the input of the amplifier from a small variable-frequency source. In essence, the power op amp then functions as an inverter, converting the energy from the DC power supply to alternating current. This is a very efficient means of controlling the speed of these motors. (Very little speed change can be brought about by varying the voltage or current applied to such motors.) Note that the so-called universal motor, although commonly powered by AC, would not undergo speed variation in this circuit—in such motors, speed is essentially constant with respect to frequency. (Also, there would be no point in using the universal motor in this arrangement because this type of motor will run on DC.)

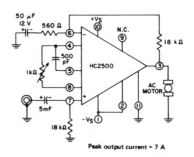

(A) Controlling speed of induction motor by frequency variation.

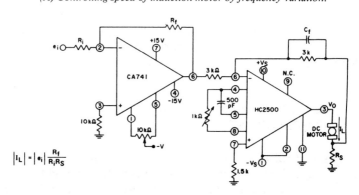

(B) Controlling and regulating current in a DC motor.

Fig. 4-10. Motor-control circuits using the HC2500 power op amp.
(Courtesy RCA)

Two things are accomplished by the scheme shown in Fig. 4-10B. The current to a DC motor is regulated so that it remains constant during load variations. This is considered desirable for various applications of powered tools; both the motor and the controlling amplifier are protected by this operational mode. Secondly, the actual level of the regulated current can be manually controlled by the DC voltage, e_i, applied to the input of the system. Inasmuch as the torque developed by a DC motor is a function of its current, this motor-control technique is best described as torque control and regulation. In practice, it will be found that the motor must readjust its speed to meet a given torque command. Therefore, such a control method may serve as a speed control for certain purposes. There will be no speed regulation, however. Indeed, speed will change more as a function of load variation than would be the case without current (torque) stabilization.

Permanent-magnet motors work well in this setup. The series motor will work too, but it will be deprived of its natural ability to meet high torque demand by drawing high current. Current-sensing resistor R_s should be as high as possible without degrading motor performance. It will generally be in the fractional-ohm region. C_f is also experimental; it protects the op amp from the consequences of abrupt load changes where the motor inductance permits development of high voltage. Essentially, this capacitor limits the transient response of the amplifier.

SCR SPEED-CONTROL CIRCUITS FOR SERIES AND UNIVERSAL MOTORS

For a number of years, the technical literature has abounded with simple SCR circuits for providing speed control of small series-type and universal motors. These circuits take the general form of the two speed-control circuits shown in Fig. 4-11. At first glance, the circuits appear similar enough to provide similar performance. This is not quite true, however. The circuit in Fig. 4-11A is not unlike the popular light-dimmer circuits, and would serve such a purpose if a lamp were substituted for the motor. However, substituting a lamp for the motor in the circuit of Fig. 4-11B would not produce the same circuitry action as the motor. Although there is no obvious feedback path in circuit (B), speed regulation occurs with the motor in the circuit. If the two circuits were compared, it would be seen that the loss in motor speed is much less in circuit (B) than in circuit (A) when equal mechanical loads are imparted to the motors, assuming the important motor-operating parameters were initially the same. In both circuits, a heat-sink should be provided for the SCR.

Because of the nature of an SCR, the motor in circuit (B) can receive current for no longer than a half cycle, or 180 degrees. Because of the phase-control provision, the duration of the current pulse delivered by the SCR will generally be less than this. However, while the SCR is blocking the passage of motor current, the motor is turning and is therefore simultaneously behaving as a generator. The topography of circuit (B) is such that the counter emf the spinning motor develops because of its generator action opposes the gate-cathode voltage of the SCR. If the added load tends to slow the motor down, this self-gener-

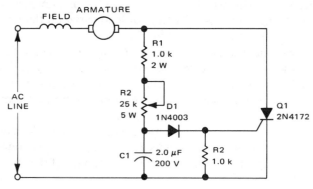

(A) Speed control without regulation.

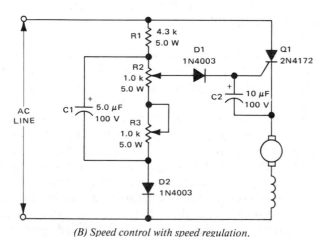

(B) Speed control with speed regulation.

Fig. 4-11. Half-wave phase-control circuits for series and universal motors. (Courtesy Motorola Semiconductor Products, Inc.)

ated voltage will decrease, allowing the SCR to trigger *earlier* in the AC cycle, thereby delivering more current to the motor than would otherwise have been the case. The added current manifests itself as electromagnetic torque, tending to speed the motor up. The closeness of the regulation depends considerably on the motor, but the fact that corrective feedback is involved is evident.

These circuits were designed for 35-volt DC series motors with 1/15 HP ratings. The circuits can be scaled up and down to accommodate other motors. In any event, it makes sense to know what kind of performance to expect. For some applications, the inherently poor speed regulation of the series motor is suitable, or even desirable. In other applications, a more nearly constant-speed characteristic is best. In the latter case, the invisible feedback path in circuit (B) electronically modifies the natural behavior of the series or universal-type motor.

When either circuit is used with a universal motor, it sometimes proves useful to connect a spst switch across the cathode-anode terminals of the SCR. Then, extra performance is forthcoming from the motor when the switch is closed. This "jump" is not as abrupt as it might initially appear; if the SCR is conducting very close to a half cycle, the closure of this switch will increase RMS motor current by approximately 40%. The control can be made idiot-proof by mechanically linking the switch to the phase-control potentiometer.

SIMPLE CIRCUIT SPEED CONTROL AND REGULATION

A simple circuit for imparting both speed control and speed regulation to small universal-type motors is shown in Fig. 4-12. Speed variation is produced by changing the triggering time of the SCR, causing the motor to receive varying portions of half cycles of the applied AC. This is somewhat reminiscent of phase-controlled light dimmers, but it will be noted that there is no phase-shift network. Even more intriguing is the lack of any obvious feedback path. (Regulators generally have sensor circuits, error amplifiers, and well-defined feedback paths.) It will be interesting to see how this control and regulating scheme operates, for its simplistic format belies its capabilities.

First, consider the nature of the universal motor; it is essentially a series motor, differing only in minor details from true DC series motors. When this motor is rotating, it generates a counter emf. In the circuit, it not only generates the counter emf when the half-wave pulses from the SCR are applied to it, but it continues to generate this counter emf during the nonconducting half cycle of the SCR. Secondly, consider the triggering characteristic of the SCR—the greater its anode-cathode voltage at a given time, the less gate voltage it needs for triggering. Combining these phenomena leads to the explanation of actual circuit operation.

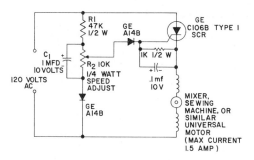

Fig. 4-12. SCR half-wave speed-control with speed regulation.
(Courtesy General Electric Company)

Assume the motor is running at some speed determined by the setting of potentiometer R2. If, for some reason, the mechanical load on the motor is increased, the motor will try to slow down; this is the natural operating behavior of series and universal motors. By slowing down, the counter emf decreases, allowing the motor to consume more current and thereby develop the higher torque needed to accommodate the greater load. However, in this circuit, when the motor slows down the reduced counter emf results in a higher *net* voltage across the anode-cathode terminals of the SCR. This is so because this net voltage is determined by the balance between the voltage impressed from the power line, and the voltage from the generator action of the motor (the counter emf). The SCR is now able to fire with *less* gate voltage than it required before the motor slowed down. This is tantamount to saying that the SCR triggers *earlier* in the AC cycle, thereby delivering increased voltage to the motor, speeding it up again. The overall action is one of *regulation* in which the motor's tendency to drop speed is largely corrected. The opposite sequence of events occurs if the motor tries to gain speed. The net result is that the motor, which by itself has very sloppy speed behavior, is electronically converted to a constant-speed machine.

The components designated in Fig. 4-12 are from an early design. Other parts and devices can be substituted, but best low-speed control will result from the use of a sensitive-gate SCR. This type of trigger control is limited to an effective phase-delay of 90 degrees, so there is inherently less low-speed control than in SCR circuits using RC phase-shift networks, as in light dimmers, which are also used for motor control (but without regulation).

AC TACHOMETER FUNCTION VIA SPECIAL OPTOCOUPLER DEVICES

The speed of a DC motor may be nicely controlled by means of a phase-locked loop. In such a system, the feedback signal must be AC and its frequency must be proportional to motor speed. The idea of using a slotted disk in conjunction with an optical-electronic arrangement is an old one. However, the implementation of such a scheme has generally involved a few headaches because of the mechanical details that had to be taken into account. In order to circumvent this, TRW and other firms have made available a family of optocouplers, so fashioned as to facilitate such applications.

Fig. 4-13 depicts a simplified phase-locked loop system for controlling the speed of DC motors. The permanent-magnet motor is frequently used. In some applications, a shunt-wound DC motor offers the advantage that the shunt field requires much less power than does the armature, or the entire motor. If, however, control is imparted through the shunt field, it must be kept in mind that the motor speed *decreases* with increasing field current and the design must be altered accordingly. In any event, the salient feature of the special optocouplers is their physical configuration; they are two-prong forks with an optical source in one side, and a detector on the other side. The slotted disk rotates between the optical source and detector, thereby producing optical chopping. The result of this is an output of pulses with a rate proportional to speed and the number of slots in the disk.

The optical source is usually an LED or an infrared-emitting diode. The detector can be a photoresponsive semiconductor device, such as a photodiode, a phototransistor, a photo-Darlington, or a photo-FET. Other devices, such as filamentary lamps and photoconductive cells have been used in the past, but they tend to introduce problems with reliability, temperature dependence, and frequency response. Other schemes can produce analogous results; for example, magnetic reluctance and magnetic induction versions of the optical chopper are feasible too. The new optocoupler devices offer the advantage that fine speed control resolution is easily realized by merely introducing a number of slots on the disk. This can be done in conjunction with the insertion of a programmable frequency divider in the feedback loop.

These optocoupler devices are known as slotted optical switches. Fig. 4-14 depicts one of the many such slotted optical switches made by TRW. This, the OPB837, features convenient mounting tabs, and utilizes a photodiode as infrared light detector. As can be seen, it is only necessary to insert an encoder disk in the slot to produce AC tachometer action, i.e., to obtain a pulse train with a repetition rate proportional to speed. The disk may have a single hole for basic operation, or it may have many holes in order to produce high resolution in an appropriate system design. Other detector schemes used in TRW slotted optical switches are shown in Fig. 4-15. Circuits in Fig. 4-15C and D contain linear amplifiers and Schmitt triggers; these circuits are available to provide either buffer, or inverter action. The circuit in Fig. 4-15B can be used to provide a unique function, which will now be described.

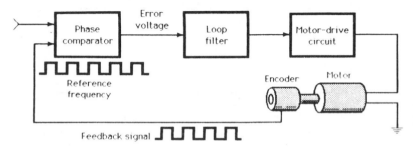

(A) Simple phase-locked loop system.

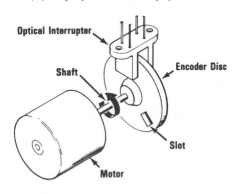

(B) Implementation of the special optocoupler, known as a slotted optical switch.

Fig. 4-13. Use of an optocoupler in a phase-locked loop motor system.

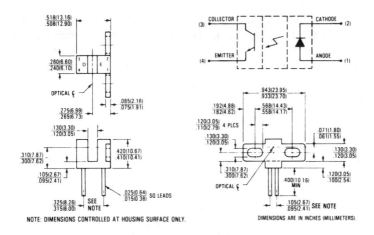

Fig. 4-14. One of the many slotted optical switches made by TRW.
(Courtesy TRW, Inc.)

The circuit of Fig. 4-15B comprises a pair of infrared-emitting diodes and their phototransistor detectors. Thus, there are two aperture-systems in the fork of the optocoupler device. In the TRW OPB822 family of optical interrupters, the aperture pairs are arranged side by side. This can be clearly seen in the drawings of Fig. 4-16. The basic idea of such an arrangement is that the chopped signals produced by the dual detectors can be used with appropriate logic circuitry to indicate the direction of rotation of the encoder disk. The logic circuit for accomplishing this is shown in Fig. 4-17.

The outputs of the two detectors can be considered as channels A and B in conformity with Fig. 4-17. The output of channel A is voltage-amplified by transistor Q1, and imparted TTL compatible rise and fall times by the Schmitt trigger designated L1. Similar action occurs in channel B via Q2 and L2. Observe that the channel A Schmitt trigger clocks the latch and the channel B Schmitt trigger connects to the D input of the latch. With this setup, the logic state of the latch's Q output tells the direction of rotation of the encoder disk. Let us see how this comes about.

Suppose the logic state of the Q output of the SN7474 is high. For this to be so, channel B had to be turned on prior to the turn-on of channel A. In other words, when the edge-triggered latch was clocked, there

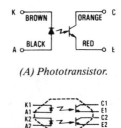

(A) Phototransistor.

(B) Dual channel phototransistor.

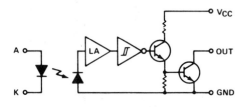

(C) Open collector logic-compatible output.

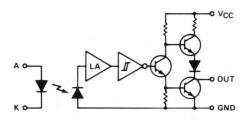

(D) Totem pole logic-compatible output.

Fig. 4-15. Additional detector arrangements in slotted optical-switch family. (Courtesy TRW, Inc.)

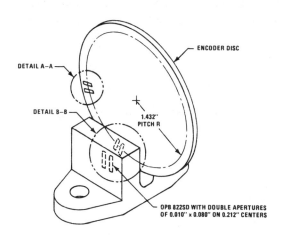

(A) Setup with encoder disk showing double aperture system.

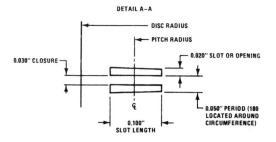

(B) Optimum slot dimensions and spacing in encoder disk.

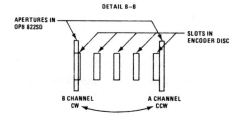

(C) Relationship of slots to aperture.

Fig. 4-16. Two-channel optical interrupter for determining direction of rotation. (Courtesy TRW, Inc.)

already was a logic-high signal at its D input. Under such conditions, the truth table of the latch would show that its Q output would go high. Referring to Fig. 4-16C, it can be seen that counterclockwise rotation exists for the condition that channel B turns on prior to channel A. The opposite conditions prevail for clockwise rotation and the Q output of the latch is then low. As shown in Fig. 4-17, an LED indicates rotational direction, but the output from Q of the latch could be used to inititate some function in the mechanical system associated with the motor.

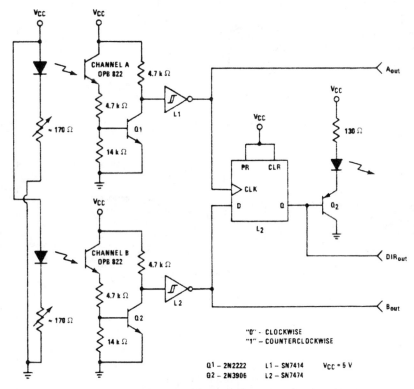

Fig. 4-17. Circuitry for use with dual-channel optical interrupter.
(Courtesy TRW, Inc.)

LOGIC-ACTUATED SWITCH PROVIDES ISOLATION

Photocouplers or photon-coupled isolators are commonly thought of as associations of light-emitting devices and photodiodes or phototransistors as detectors. Unique components incorporating photon isolation, but utilizing a light-controlled thyristor are made by General Electric Company. The thyristor acts as a gateless triac, or as a diac, in the sense that it provides bilateral conduction when turned on by the AC applied to it through a load circuit. This characteristic makes it useful as a trigger source for an external triac, which may, in turn, act as a switch for a large load, such as an induction motor.

Just such an application is shown in Fig. 4-18. With a small parts count, the features of electrical isolation and receptivity to logic-level control are achieved. The basic purpose for using this scheme would be to avoid the contact and mechanical problems encountered when using electromagnetic devices (contactors and relays) for switching multiampere inductive loads. The circuit is also useful for turning on and off resistive loads such as heaters and lights. In all cases, consideration should be allowed for inrush currents—stationary motors, inactive heaters, and cold filaments in lamps all consume surge currents when first turned on. That is why a fairly large triac, the SC16OD is indicated in Fig. 4-18.

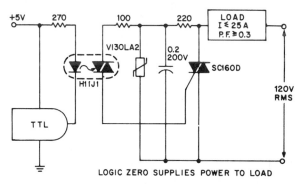

Fig. 4-18. Logic-actuated contactor suitable for AC induction motors.
(Courtesy General Electric Company)

The V13OLA2 is a metal-oxide varistor. This General Electric device absorbs the energy of voltage transients. It protects the thyristor element in the J11J1 and also prevents accidental turn-on of the triac from line transients. It also acts to reduce electromagnetic interference when the triac is actuated. A practical hint involves the connections of unused pins on the 6-pin dual in-line package of the H11J1. Whereas pin 3 is truly a blank pin, pin 5 is internally connected to the substrate of the photothyristor. Because of this, pin 5 should be left floating—it must not be grounded or either purposely or inadvertently connected to any part of the external circuit. Otherwise, the photothyristor will not be properly responsive to light actuation.

ELECTRONIC SWITCH FOR STARTING SINGLE-PHASE INDUCTION MOTORS

The single-phase induction motor is a wonderful machine in that it is inherently rugged and is easy to manufacture. It has no brushes, commutators, slip rings, or other sliding contacts. Its electrical and mechanical characteristics are about what the doctor ordered for a wide variety of tasks, especially around the home. But, unfortunately, it develops no starting torque in its basic form—something has to be done to initially get it going. Once brought up near its normal running speed, the starting mechanism, whatever its nature, can be dispensed with. Normal running speed is something close to synchronous speed. For the commonly encountered 60-Hz four-pole types, this is slightly less than 1800 rpm. In 60-Hz two-pole induction motors, this is slightly under 3600 rpm.

A popular means of making these motors strongly self-starting is to design them with an extra stator winding and to impress this starting winding with a current 90 degrees out of phase with that applied to the main or running stator winding. The phase displacement is produced by a large electrolytic capacitor connected in series with the starting winding. Then, after the motor has accelerated up to perhaps 80% or so of its normal running speed, a centrifugal switch disconnects the starting winding and the motor is able to rely on its own torque to bring it to running speed. As long as the starting winding, with its capacitor, is in the circuit, the motor behaves essentially as a two-phase machine. (The two-phase induction motor develops a rotating field, and therefore is self-starting. Three-phase motors, for the same reason, develop high torque at standstill, and are self-starting without any switching mechanism.)

The trouble with centrifugal switches and electrolytic capacitors is that they are both relatively high maintenance items. When the single-phase induction motor gives trouble, the source of the malperformance is rarely in the motor itself, unless conditions have become sufficiently abusive to burn out the stator winding(s).

The electronic-switching scheme shown in the motor-start circuit of Fig. 4-19 provides a partial solution to this dilemma. Although it retains the electrolytic capacitor, it eliminates the centrifugal switch. This is a good remedy, inasmuch as it is much easier to correct for a bad capacitor than it is to overhaul or replace a defunct centrifugal switch. The operation of this circuit is predicated upon the inrush current of a motor starting up and gathering speed. At first the inrush current is high and sufficient voltage drop is developed across R1 to keep the triac triggered. This, in turn, keeps the phase-shifting capacitor, C2, in the starting-winding circuit. When the motor has attained sufficient speed to be on its own, the decreased current through R1 will no longer develop sufficient voltage drop to trigger the triac, thereby opening the starting-winding circuit. Obviously a bit of experimentation is in order for the optimum value of R1. If this resistance is too high, the starting circuit will not be opened during normal running and both the motor and the triac will overheat.

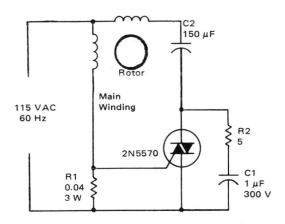

Fig. 4-19. Electronic switch for capacitor-start induction motors.
(Courtesy Motorola Semiconductor Products, Inc.)

TRIAC PHASE-CONTROL CIRCUIT FOR SPEED ADJUSTMENT

A triac tends to be a more satisfactory device for speed control of universal motors than ordinary SCRs. Although both devices utilize phase control to vary the current through the motor, the triac, being a full-wave device, symmetrically controls the phase of *both* half cycles of the applied AC. The resultant full-wave current format then produces smoother motor operation than can readily be attained from the half-wave rectification of SCRs. This tends to be particularly noticeable at low speeds. The triac phase-control circuit shown in Fig. 4-20 is depicted for a family of 12-ampere triacs. However, the same component values generally suffice for both smaller and larger devices as well.

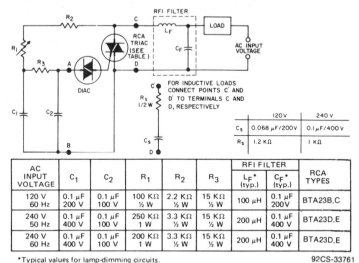

AC INPUT VOLTAGE	C_1	C_2	R_1	R_2	R_3	RFI FILTER		RCA TYPES
						L_F* (typ.)	C_F* (typ.)	
120 V 60 Hz	0.1 µF 200 V	0.1 µF 100 V	100 KΩ ½ W	2.2 KΩ ½ W	15 KΩ ½ W	100 µH	0.1 µF 200 V	BTA23B,C
240 V 50 Hz	0.1 µF 400 V	0.1 µF 100 V	250 KΩ 1 W	3.3 KΩ ½ W	15 KΩ ½ W	200 µH	0.1 µF 400 V	BTA23D,E
240 V 60 Hz	0.1 µF 400 V	0.1 µF 100 V	200 KΩ 1 W	3.3 KΩ ½ W	15 KΩ ½ W	200 µH	0.1 µF 400 V	BTA23D,E

*Typical values for lamp-dimming circuits.   92CS-33761

Fig. 4-20. Triac phase-control circuit for adjusting speed of universal motors. (Courtesy RCA)

A generalized load is indicated in Fig. 4-20 rather than a motor symbol because this circuit is also widely used for light dimmers and for the control of heaters. Often, a DC series motor will also work, but efficiency and commutation will not be as good as with the universal motor.

This circuit features a double time constant phase-shift network. As with SCRs, this reduces hysteresis in the firing of the triac, thereby making the manual adjustment of speed more repeatable. (This control technique provides for adjustment of the motor's speed, but does not produce speed regulation.)

Adding to the reliability of this circuit is the diac trigger device. This is essentially a gateless thyristor designed to break down on both polarities of the AC wave. A given diac can properly trigger a wide variety of thyristors. The elemental low-pass filter comprised of L_F and C_F attenuates much of the radio-frequency interference that would otherwise get back to the power line. Such high-frequency energy is generated by the extremely rapid turn-on time of the triac. Radiation from the power line can cause much trouble.

From the table in the schematic diagram, it can be seen that this 12-ampere family of triacs includes devices rated for 240-volt as well as 120-volt operation. These are, however, intended for 50-Hz and 60-Hz service. Special triacs are marketed for operation from 400-Hz power sources; triacs for even higher frequencies have been specified for special applications, but are not readily available. The 400-Hz phase-controlled circuit shown in Fig. 4-21 uses triacs specified for this power-line frequency. This circuit uses different RC values in the phasing network, but is otherwise similar to the 60-Hz circuit. The triac is of the T41113, T41114, T41115 series.

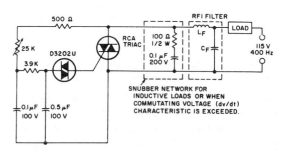

Fig. 4-21. 400-Hz, phase-controlled triac circuit. (Courtesy RCA)

TRANSISTORS AS MOTOR-CONTROL DEVICES

A pause is in order to contemplate the status of transistors as power-control devices for motors. Extensive applications were visualized when the power transistor made its debut. However, early transistors did not survive the harsh nature of motor loads too well.

Motor-control techniques have generally been dominated by thyristor power devices, notably the SCR and the triac. There have been exceptions to this, depending upon the size of the motor and whether it was an AC or DC type. However, for motor sizes beyond about the one-tenth horsepower rating, bipolar transistors have been at a disadvantage. They tended to lack the voltage and current ratings required, and were further handicapped by their one-polarity operational requirement. Of course, various strategems can be used to tolerate these shortcomings, such as paralleling, the use of diodes, and novel circuit arrangements. The modern trend still makes use of transistor power devices, but they now assume the following more useful formats:

- Bipolar transistors are now available with much better ratings—these include current, voltage, SOA, and thermal resistance. Garden-variety power transistors of the past are overshadowed by the newer versions for motor-control applications.
- The bipolar Darlington transistor, for many years a lackluster power device, has become a capable brute-force element with potentialities for controlling both fractional and integral-horsepower electric motors. Further facilitating practical application, the Darlington device is generally easier to parallel than single-transistor types. These rugged Darlingtons still retain their easy-drive characteristics because of their relatively high beta ratings.
- Both the modern single-element bipolar transistor and the newer Darlingtons are available in PNP as well as NPN polarities. Often, there is a very close match between NPN and PNP counterparts. This simplifies the design and implementation of push-pull and bridge power amplifiers; it also helps provide for regenerative or dynamic braking and allows for more straightforward design of polyphase circuits. Complementary-symmetry power-amplifier circuitry making use of NPN–PNP devices usually features a relatively low parts count. Thus, the traditional concept that the PNP format was unsuitable for high-power devices is no longer valid.
- Power MOSFET devices have progressed considerably since their commercial debut. At first a novelty, these devices now successfully compete with conventional bipolar transistors, with power Darlingtons, and with thyristors. They have intrinsically fast operating characteristics, and tend to be immune to thermal runaway. For certain motor-control applications, these devices possess the unique feature that they can operate with either polarity. Even more design flexibility results from the fact that both N-channel and P-channel types are available. (This corresponds respectively to NPN and PNP bipolar types.)

MOTOR CONTROL WITH A DUAL POWER OP AMP

The L272 is a dual op amp with power capability. Whereas commonly used op amps, such as the popular 741, generally cannot safely provide more than 100 milliamperes, the output-current capability of each of the L272 op amps is one ampere. Inasmuch as these devices can also operate from a 24-volt source, applications to the control of small motors are feasible. Such control is facilitated by the differential input and the push-pull output configurations. As with other ICs, internal circuitry is direct-coupled. This enables easy application to permanent magnet, series, shunt, and other DC motors. However, series DC motors and universal motors will not reverse their direction in the illustrated circuits. (In such motors, direction of rotation is not changed by polarity reversal at the terminals. To change direction, either the series-field connections, *or* the armature connections, but *not* both, must be reversed.)

The circuit shown in Fig. 4-22A enables control of direction and speed by means of an analog control voltage. This neatly circumvents the need for a dual power supply which would be needed if only a single op amp were used. A nice feature of this arrangement is that the control voltage for zero speed (standstill) tends to be the same regardless of the main supply voltage. Thus, for many purposes, the system can be operated from a simple, nonregulated power supply.

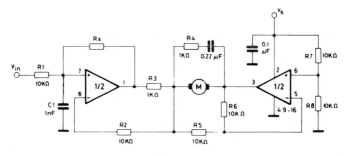

(A) Speed control from a variable voltage source.

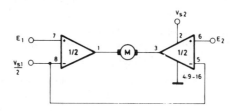

(B) Direction selection from a microprocessor or other digital source.

Fig. 4-22. Bidirectional motor control with the L272 power op amp.
(Courtesy SCS Semiconductor Corp.)

The circuit arrangement of Fig. 4-22B provides for logic-controlled direction of rotation. This is a good peripheral output interface for certain microprocessor systems where it is desired to select any of three motor-operating conditions—clockwise or counterclockwise rotation, or standstill. Standstill occurs when both input terminals are impressed with like logic levels. Unlike logic signals at the inputs cause rotation in one direction or the other. Thus, the arrangement functions like a single-pole double-throw switch from the viewpoint of the motor. In order for this scheme to be operative, the main supply voltage, V_{S2}, must be greater than the logic supply voltage, V_{S1}. Inasmuch as the logic supply voltage will usually be 5 volts, this requirement is readily met in practice. Note that the logic voltage levels are applied to terminals 6 and 7. The $V_{S1}/2$ voltage applied to terminal 8 sets the standstill condition of the motor.

FULL-WAVE MOTOR CONTROLLER USING ANTIPARALLEL SCRS

The circuit of Fig. 4-23 may be said to be a classic approach to full-wave phase control of a motor. The implication is that the simplest devices are used, these being SCRs, transistors, and diodes—there are no triacs, diacs, or unijunction transistors. And of the devices used, only the SCRs carry the motor current. Some designers opt for this approach on grounds that one can anticipate greater reliability, especially where abusive overload conditions are encountered. In some measure, such a contention is subjective, evidencing the biases of the designer. To give credit where due, however, the SCR tends to be a more rugged device than the triac. If need be, one can procure SCRs with voltage, current, frequency, and peak-current ratings far beyond what is available in triacs. Also, in some instances, commutation problems in triacs are less likely to find counterpart malperformance with SCRs. Perhaps the circuit of Fig. 4-23 would not appeal to the parts-count conscious constructor; yet, the circuit would hardly qualify as a complicated one.

It can be seen that the SCRs are connected in an antiparallel mode, or back-to-back. This enables the pair to simulate the operation of a triac; indeed, explanations of triac behavior often make use of the back-to-back SCR analogy. The overall result of such an analogous circuit is reasonably close. A practical difference is that the triac enjoys the luxury of simple triggering via a single gate, whereas the SCR duo requires individual triggering of the separate gates. This is remedied

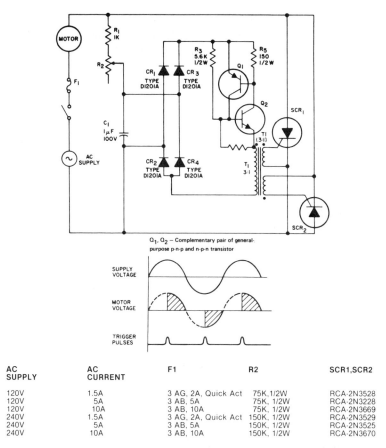

Fig. 4-23. Full-wave motor control circuit using antiparallel connected SCRs. (Courtesy RCA)

easily enough by the use of a pulse transformer with two secondaries. With this provision, the primary of the transformer looks like the gate of the triac. It will be noted that the secondary windings of the pulse transformer are phased to forward-bias both gates simultaneously. However, *only* that SCR in which the anode is positive relative to the cathode fires. This implies that sequential trigger pulses produce alternate firing of the SCRs. A little thought experimentation with this principle will reveal that a single secondary winding would not work, nor would a center-tapped winding be feasible—two conductively isolated windings are at the heart of this scheme.

Another interesting circuitry adaption is found in the trigger generator comprising complementary bipolar transistors. This exact circuit is found in the technical literature dealing with the analogous circuit of the SCR; it is also used in explaining such negative-resistance devices such as the SGS, the PUT, the diac, and the UJT. The salient feature of this circuit is that it is regenerative—the collector of each transistor feeds the base of the other.

Suppose that both transistors are initially in their nonconductive state. If a pulse of forward bias is impressed at one base, an amplified pulse of forward bias appears at the other base. In turn, the first base receives reinforced forward bias, etc. The circuit experiences a cumulative sequence of actions leading to a locked condition with both transistors heavily saturated in their conductive states. Because this process is regenerative, the transition from the off to the on state is almost instantaneous, and is accompanied by a sharp pulse which is suitable for triggering SCRs. It is not even necessary to inititate the action with a pulse—a slowly rising signal level suffices. In the actual circuit of Fig. 4-23, the slowly rising (relatively speaking) base signal is obtained from the RC phase-shift network comprising R1, R2, and C1. By adjusting R2, the phase of the voltage derived from this network can be varied, thereby controlling the time of triggering of the SCRs, which, in turn, determines the effective value of the voltage applied to the motor.

This system is best suited to the control of large universal motors. Series-type DC motors can sometimes be successfully used up to 60 Hz. At higher frequencies, there would be high losses and also commutation problems. Permanent-magnet and shunt motors cannot be operated from sources providing AC waveforms such as this controller provides. A split-phase induction motor of the permanent-capacitor type can, surprisingly, undergo considerable speed variation when energized from this controller; this is particularly true when the motor load is a fan. (It is also true that fan performance changes greatly for relatively small speed changes.) The experimenter may wish to use a controller of this type with shaded-pole induction motors, although these are not often encountered in large sizes due to their inherently low starting torque.

Almost any NPN–PNP small-signal transistors will serve in the trigger circuit and little is gained by striving for a close match in transistor parameters. However, it is often wise to use overrated components in industrial equipment where RFI, EMI, and transients tend to be much stronger than is generally encountered in communications or instrumentation circuits. Thus, a pair of transistors with higher voltage and current capability than is ordinarily identified with small transistors would be suitable for this controller; such a pair of transistors is the RCA1A18 NPN type, and the RCA1A19 PNP type. These devices have one-ampere maximum current ratings and are available in the JEDEC TO-39 package. Heat sinking is not necessary because of the relatively low dissipation incurred in the regenerative switching process.

There is no performance penalty associated with this trigger circuit. The controller provides a phase adjustment range of 5 to 170 degrees; for most practical purposes, this may be said to correspond to a power control range of zero to 100%, where 100% is the power that the motor would extract directly from the utility line.

SPEED CONTROL OF LARGER INDUCTION MOTORS

The motor-control techniques hitherto discussed pertain to motors in which speed can be controlled or stabilizied by varying the voltage or current applied to them. This has included nearly every type of motor except ordinary induction motors. In some instances, very small induction motors with wound rotors, split-phase types (especially the permanent capacitor split-phase motor), and shaded-pole types can be controlled by voltage; the speed variation is relatively limited and is best realized when the load is a fan or blower. Textbooks dealing with induction motors often ignore these exceptions, simply labeling induction motors as constant-speed devices. Of course, the induction motor does shed speed with load, but the implication is that the slip or departure from synchronous speed (the rotational speed of the magnetic field) is only a few percent, and moreover is *not* primarily influenced by voltage or current.

The induction motor is the workhorse of industry, especially in its three-phase format. The three-phase induction motor is self-starting, very efficient, easily reversed, electrically simple and mechanically rugged, and displays desirable torque and speed characteristics for many applications. A tempered statement applies to single-phase induction motors with the major exception being that they are not inherently self-starting. Although there is no magic line of demarcation, three-phase motors are largely found in the integral-horsepower ranges. Single-phase induction motors for heavy-duty uses feature cast rotors (as do the three-phase types) and have fractional-horsepower ratings from about one tenth to one horsepower. Such ordinary induction motors are widely used but are not amenable to speed control by any of the techniques thus far described.

The induction motors that can be controlled, at least to some extent, by some of the hitherto described circuits are generally smaller than one tenth horsepower and are of the specialized types alluded to above. Control should only be attempted with circuitry supplying full-wave AC waveforms to the motor.

In the past, the only practical (albeit expensive) way to vary the speed of the three-phase or ordinary single-phase motor was via an electromechanical frequency changer—this generally comprised an alternator driven either by an internal-combustion engine, or by a DC motor. With the advent of solid-state power devices, a new approach has seen considerable development. It remains understood that the supply frequency must be controlled, but this is now done by means of electronic inverters. This has endowed industrial processes with the basic desirable features of induction motors with the added advantage that these have, in essence, become variable-speed machines. The initial investment is a fraction of the cost of the mechanical approach. Moreover, the electronic inverter and induction motor combination is admirably suited for traction vehicles, where the maintenance of brush-commutator machines can be entirely circumvented.

In the technical literature dealing with the control of large induction motors, it is commonplace to use block diagrams, functional schematics, and simplified and partial circuits more extensively than is customary with applications appealing more to hobbyists and experimenters. It is assumed that control of larger motors is more the domain of professional designers and the important thing to convey is the basic principle, rather than details pertaining to parts, devices, and localized circuits. These motor controllers tend to be considerably more complex than those that have been discussed for smaller motors. Then, too, we definitely find ourselves involved in a hazardous voltage-current region where the average electronics practitioner's work habits are not always compatible with safety of person and equipment. And, finally, this is a subject unto itself—a specialty requiring voluminous treatment which is not within the scope of this book. However, a brief allusion will be made to the important aspects of large induction motor control because of the necessarily nebulous distinction between large and small induction motors.

To start with, it should be realized that the average 60-Hz induction motor is inherently capable of running at a fraction of its rated speed, and also at several times its rated speed. To be sure, special motors can be designed for such purposes that will do the job somewhat more efficiently, and possibly smoother at low speeds than garden-variety types.

However, both practical experience and mathematical design show that ordinary motors have considerable capability for such service. This tends to be true even in many instances wherein the applied waveform is closer to square than sinusoidal, although departure from sine-wave excitation can degrade operating efficiency and can roughen the torque characteristic under certain speed/load conditions. The three-phase machine, especially, is quite forgiving; like a multicylinder engine, its inherent smoothness overcomes nonideal conditions.

Notwithstanding, there is *one* criterion that must be met with all variable-frequency control systems for induction motors. Except if one is content with a small range of speed control, *the applied voltage to the motor should vary* (either manually or automatically) *directly as the frequency.* Thus, if the frequency (and speed) is doubled, twice the rated (60-Hz) voltage should be impressed on the motor. Conversely, if the frequency is halved to 30 Hz in order to bring about half-speed operation, the motor should receive one half of its 60-Hz voltage. If this were not done, the motor would suffer severe loss of torque at the higher frequency, and would be severely overloaded electrically at the lower frequency. Note, however, that the change in voltage, as such, is *not* the speed-changing agency.

The block diagram of Fig. 4-24 depicts the general control scheme used with three-phase induction motors. It is basically a driven inverter using two output stages per phase. These output stages function as switches and can be sequenced in various ways for powering the motor. Usually, it is desirable to make the best use of both digital and analog operation. For power-handling efficiency, the output stages should have only two states—on and off. For good torque characteristics, it is best that the motor current be sinusoidal. Although wave purity is not as important as in audio amplifiers, a reasonable approximation to a sine wave results in smooth motor operation and prevents unnecessary temperature rise from eddy currents and hysteresis. Two methods of complying with the needs of the output stages and the motor have become popular. These are the six-step system and the pulse-width modulation system. Both represent formats for switching the output stages so that three-phase quasi-sine waves will be delivered to the motor. Both systems make use of motor inductance to smooth the pulsed waveforms for a better approach to a sinusoidal shape. With small variations, both methods can be functionally represented by the general setup shown in Fig. 4-24. (To the right of the dashed line, the system is substantially the same for both methods. This, it will be seen, represents the power-driven circuitry.) The system, as shown, is set up for the six-step mode.

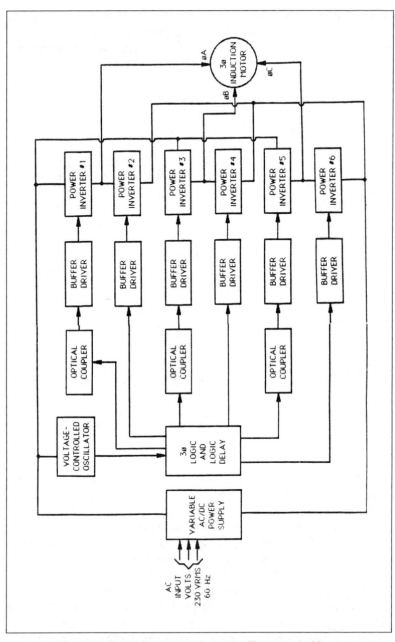

Fig. 4-24. Generalized setup for controlling speed of large induction motors.

As its name implies, the six-step method fabricates quasi-sine waves from six basic constructs—square waves with predetermined amplitudes, durations, and times of occurrence. Fig. 4-25 illustrates such a three-phase format of voltages and currents. Inasmuch as it is the current that develops the torque in the motor, particular attention is directed to its approximation to sinusoidal shape. The three top waveforms are phase voltages. The three lower waveforms are phase currents. In this regard, the dashed waves in Fig. 4-25 represent the motor current and show the smoothening effect of motor inductance.

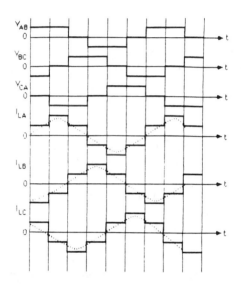

Fig. 4-25. Phase waveforms in six-step, three-phase motor-drive system.

The pulse-width modulation system compares a low-frequency sine wave with a high-frequency triangular wave, obtaining a wavetrain of varying duty-cycle square waves. This, again, is a quasi-sine wave insofar as concerns its circuit effects. In Fig. 4-26A the manner of bringing about this result is illustrated; either a voltage comparator, or an overdriven op amp work well in this capacity. The wavetrain in Fig. 4-26B represents one of the motor phase currents, and the dashed wave shows the smoothening effect of motor inductance. As with the six-step method, the best features of digital and analog technology are systematically deployed—the output stages perform as switches, and the motor receives at least quasi-sine waves of drive currents.

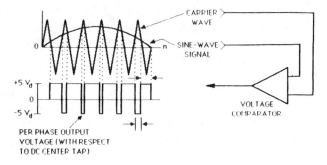

(A) High-frequency triangular wave (carrier) compared with low-frequency sine wave produces a pulse-width modulated wave.

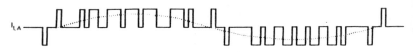

(B) Phase-current wave in motor simulates original sine wave through agency of varying-duration pulses.

Fig. 4-26. Phase waveforms in pulse-width modulation motor-drive system.

More-detailed circuitry of a drive system representative of the six-step wave-synthesis technique is shown in Fig. 4-27. In this case, the design is intended to control a 2-HP 230-volt, three-phase induction motor. As has been previously inferred, six-step systems have much in common with pulse-width modulation systems and the power-function blocks of the two systems tend to be readily interchangeable. This being the case, it might prove profitable to make the necessary minor changes and adaptations so that Fig. 4-27 can handle pulse-width modulated, rather than six-step, signals. It is easier to obtain smooth motor operation, especially at low speeds, with pulse-width modulation. Also, the need to vary the DC voltage supplied to the inverter can be dispensed with—although motor voltage must still be varied with speed, this can be achieved by varying the width of the pulses. And best of all, the semiconductor firms have made available ICs for attaining all of the logic required in a three-phase pulse-width modulation system with minimal complexity and expense. This has come about as the result of these companies' previous experience with pulse-width modulation ICs used in switching power supplies.

In dealing with high power levels and large induction motors, it is often desirable that the inverter be current, rather than voltage, fed. That

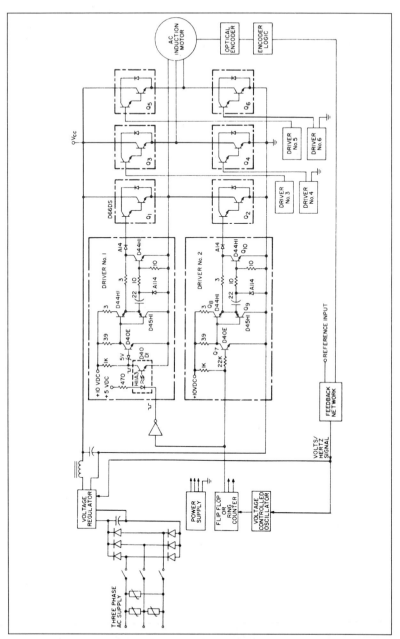

Fig. 4-27. Detailed circuitry of six-step drive system suitable for a two-horsepower induction motor. (Courtesy General Electric Company)

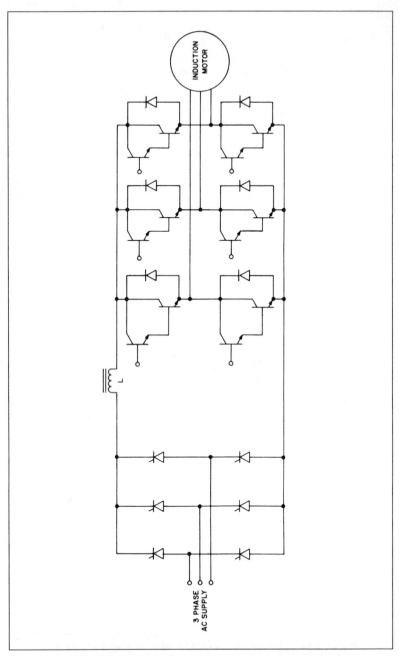

Fig. 4-28. Basic concept of the current-fed inverter.

is, the output element of the power supply should be an inductor, rather than a capacitor. This may appear to be a triviality, but it tends to enhance the reliability of the overall system. The circuit of Fig. 4-28 shows one way of accomplishing this. The SCRs in the three-phase power supply can be phase-controlled in order to control motor current, and the six inverters can be driven either by the six-step or PWM switching formats. This technique greatly reduces the peak currents in the output stages and limits the tremendously high starting and blocked-rotor currents of large motors. Experience indicates that the current-driven inverter prolongs motor life. And, because switching transients are reduced in violence, RFI and EMI problems tend to be less severe than with voltage-driven inverters. A practical problem, however, is that at low frequencies, the inductor may have to be inordinately large.

DEDICATED IC FOR CONTROLLING FOUR-PHASE STEPPING MOTORS

Servo systems have long been used to both control and position the shafts of motors. Various degrees of success have been attained; in principle, a high amplification of the error signal should result in arbitrarily precise position control. In practice, high gain provokes problems of overshoot, instability, and oscillation. The critically damped, mathematically obedient systems described in textbooks are not found in most applications because of unpredictable, varying, and uncontrollable inertial and frictional forces from bearings, gears, shafts, belts, etc. A better way becoming popular is through the computerized control of stepping motors.

Stepping motors advance one step at a time in response to appropriate pulses. The basic idea of control is to supply predetermined sequences of pulses. If these can be controlled in rate, time of occurrence, and format of application to the windings of the stepping motor, the direction, speed, and stopping position of the motor can be precisely put through its paces. The instabilities of servo systems, such as hunting, are minimal or absent inasmuch as operation is not dependent on feedback, or amplification of an error signal. But, until recently, the electronics needed for this purpose entailed numerous discrete components in large racks and panels.

The Cybernetics Micro Systems CY525 Intelligent Ramping Stepper-Motor Controller is a 40-pin dedicated IC designed to interface a command source and a power driver for the motor. The command source can be an ASCII keyboard via high-level language, or a computer or microprocessor in which instances the commands can use binary code. The salient feature of ASCII keyboard control is that a system can be prototyped prior to introducing computer commands. It is much more convenient, and usually more economical, to explore a system's behavior by typing in commands than by harnessing the services of a computer. Of course, many systems can be successfully implemented by retaining the keyboard control.

Performance control is almost limitless and involves primarily programming skills, rather than hardware circuitry design. Up to 10,000 steps per second can be handled. Best of all, sequences of high-level commands can be stored internally, and can be executed upon command. A continuous-run mode simulates conventional motors by supplying an unlimited number of steps. When operating in this fashion, the stepping motor behaves similarly to a synchronous motor in that the rotational speed is locked to the frequency or pulse-rate of its drive source.

Fig. 4-29 depicts a simple stepping-motor control setup utilizing the CY525. The ASCII high-level commands and decimal values from the keyboard are translated by the CY525 into a logic format, which is then power boosted to control the operation of the stepping motor. Power boost is accomplished by the ULN 2068B, a power Darlington transistor array made by Sprague. Other power-transistor schemes may be used—the basic program being to process the logic output of the CY525 into stepped levels with sufficient power to drive the motor. In most cases, the 7404 hex inverter must be used to interface the CY525 with the power stages.

Inasmuch as so many functions have been incorporated on the CY525 IC, it is only natural to ponder why the power-drive circuitry wasn't also included. Not only do nasty thermal problems assert themselves when brains and brawn are integrated on a single chip, but a point is generally reached in systems where flexibility and versatility are sacrificed if the whole system resides in a single IC. It is often better to have the system divided into several subsystems. The parts count and complexity is still dramatically reduced from what it would be with discrete-device design, but the user is accorded more options for adaptation to the unique needs of his situation.

The internal clock of the CY525 can be made operational by simply connecting a 2- to 11-MHz series-resonant crystal to pins 2 and 3. A 3.58-MHz TV color-burst crystal suffices for this purpose. However, the higher the clock frequency, the higher the maximum stepping rate. An 11-MHz crystal will allow nearly 10,000 steps per second.

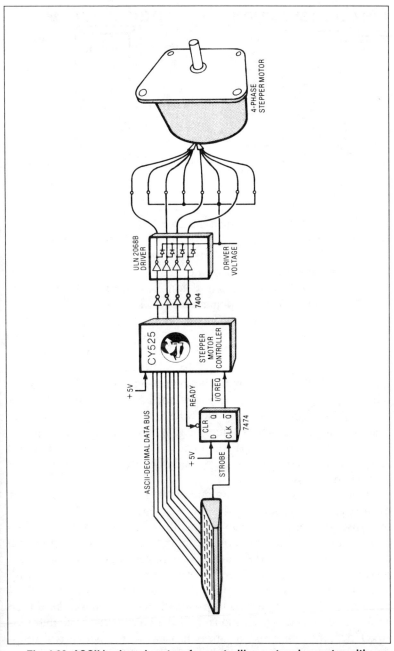

Fig. 4-29. ASCII keyboard system for controlling a stepping motor with the CY525. (Courtesy Cybernetics Micro Systems, Inc.)

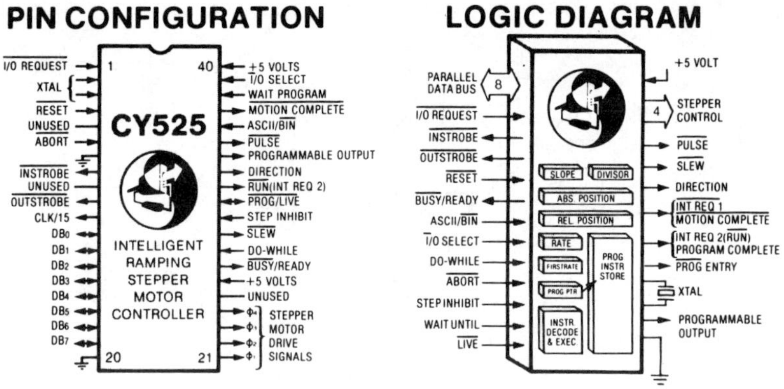

PIN CONFIGURATION

```
                  ┌────────┐
  I/O REQUEST ──→ │1     40│ ←── +5 VOLTS
         XTAL ──→ │        │ ←── I/O SELECT
        RESET ──→ │        │ ←── WAIT PROGRAM
       UNUSED ──→ │ CY525  │ ←── MOTION COMPLETE
        ABORT ──→ │        │ ←── ASCII/BIN
      INSTROBE ←─ │        │ ←── PULSE
       UNUSED    │        │ ←── PROGRAMMABLE OUTPUT
     OUTSTROBE ←─│        │ ←── DIRECTION
        CLK/15 ←─│        │ ←── RUN(INT REQ 2)
          DB0 ←─→│        │ ←── PROG/LIVE
          DB1 ←─→│INTELLIGENT│ ←── STEP INHIBIT
          DB2 ←─→│ RAMPING │ ←── SLEW
          DB3 ←─→│ STEPPER │ ←── DO-WHILE
          DB4 ←─→│  MOTOR  │ ←── BUSY/READY
          DB5 ←─→│CONTROLLER│ ←── +5 VOLTS
          DB6 ←─→│         │ ←── UNUSED
          DB7 ←─→│         │ ──→ ⎱ STEPPER
                │         │ ──→ ⎰ MOTOR
                │ 20    21│ ──→ ⎱ DRIVE
                └────────┘ ──→ ⎰ SIGNALS
```

LOGIC DIAGRAM

```
   PARALLEL
   DATA BUS ⟨8⟩                         ←── +5 VOLT
                                        
  I/O REQUEST ──→                  ⟨4⟩→ STEPPER
    INSTROBE ──→                        CONTROL
   OUTSTROBE ←──  [SLOPE][DIVISOR]  ──→ PULSE
       RESET ──→                        SLEW
   BUSY/READY ←── [ABS. POSITION]   ──→ DIRECTION
    ASCII/BIN ──→ [REL. POSITION]       INT REQ 1
   I/O SELECT ──→ [RATE]                MOTION COMPLETE
    DO-WHILE  ──→ [FIRSTRATE] PROG      INT REQ 2(RUN)
       ABORT  ──→            INSTR  ──→ PROGRAM COMPLETE
  STEP INHIBIT ──→ [PROG PTR] STORE ──→ PROG ENTRY
                                    ──→ XTAL
   WAIT UNTIL ──→ [INSTR                PROGRAMMABLE
         LIVE ──→  DECODE                 OUTPUT
                 & EXEC]
```

Fig. 4-30. Pin configuration and logic diagram of the CY525 stepping-motor controller. (Courtesy Cybernetics Micro Systems, Inc.)

The pin and logic diagrams of the CY525 are shown in Fig. 4-30. It should be obvious that the 40-pin IC comprises a very involved system, one that would dwarf the stepping motor itself if construction were attempted with discrete devices. The logic diagram also depicts in simplified form the internal architecture of the CY525. Note the PROG INSTR STORE; this enables high-level commands to be typed into memory so that they may be executed at a later time. Sixty bytes of program commands can be stored. Another salient feature of the CY525 is represented by pin 36, appropriately labeled ASCII/BIN. The logic level applied to this pin determines whether the CY525 is receptive to ASCII data or to binary arithmetic commands. Thus, a logic high enables operation from ASCII commands; a logic low makes the system responsive to binary code, such as might be received from a computer or a microprocessor.

Table 4-1 is a list of the programming features of the CY525. A study of these capabilities should drive home the fact that versatility of control greatly exceeds what can be readily accomplished with analog-servo systems. It is true that the stepping motor cannot compete with conventional motors in the brute-force parameters of torque and horsepower; within their capabilities, however, stepping motors together with logic-circuit drive systems provide superior control accuracy and repeatability. Relatively large and powerful stepping motors have been designed for special applications, and the commercial availability of larger stepping motors is now likely in view of the high-current solid-state power devices now available for pulsing these motors. Although

Table 4-1. Programming Features of the CY525 Stepping-Motor Controller (Courtesy Cybernetics Micro Systems, Inc.)

Programmable via ASCII keyboard	Two interrupt request outputs
ASCII-decimal or binary communication	Programmable output line
Single 5-volt power supply	Programmable delay
27 hi-level language commands	Verify register/buffer contents
Stored program capability (60 bytes)	Several sync inputs and outputs
Linear acceleration, definable	Abort capability
Change rates while stepping	Step inhibit operation
Read position on-the-fly	Ability to turn off phases
10,000 steps per second (11 MHz xtal)	Slewing indication output
Absolute/relative position modes	"Dowhile" and "Wait Until" commands
Define starting rate and slew rate	"Jump to" command
Ramp-up/slew/ramp-down	Loop command with repetition count
Hardware or software start/stop	Allows address labels for loop and jump commands
Software direction control	Unlimited number of steps in continuous mode

the logic instructions needed to implement the performance features listed in Table 4-1 will not be dealt with here, putting the stepping motor through its paces should prove an intriguing project for the enterprising programmer.

The command summary of the CY525 is shown in Table 4-2. Note that there are twenty-seven of these high-level instructions for bringing about the various operating modes.

A setup for placing the CY525 in operation is shown in Fig. 4-31. This is essentially a more-detailed version of Fig. 4-29. Note that LEDs are provided for visually indicating states and state transitions. The basic idea is to establish a prototype system with the aid of the ASCII keyboard. If desired thereafter, the commands can originate in an eight-bit data bus, a microcomputer, or in data stored in PROM, EPROM, or ROM. This prototyping capability greatly simplifies the tailoring of stepping-motor control to the requirements and limitations of a practical system.

Table 4-2. Command Summary of the CY525 Stepper-Motor Controller (Courtesy Cybernetics Micro Systems, Inc.)

Aa	Absolute location specified
B	Bitset (control output = 1)
C	Clearbit (control output = 0)
Dd	Delay specified milliseconds
E	Enter program into CY525
Ff	First step rate
G	Go (begin stepping)
H	Haltmode or continuous step
I	Initialize-software reset
Jj	Jump to prog buffer location
Lc,a	Loop through prog segment
Nn	Number of steps
Oo	Offset drive signals as req'd
Pp	Position to step to
Q	Quit entering program code
Rr	Rate of stepping
Ss	Slope of accel/decel
Tt	Til pin 28 HI, repeat program
U	Until WAIT LO, wait here
Vv	Verify buffer contents
W	Wait until WAIT line HI
X	Execute stored program
Zz	Divisor for slope
+	Set clockwise direction
-	Counterclockwise direction
O	Return to command mode
$	Marker to Jump & Loop to

These commands can be either stored or executed.

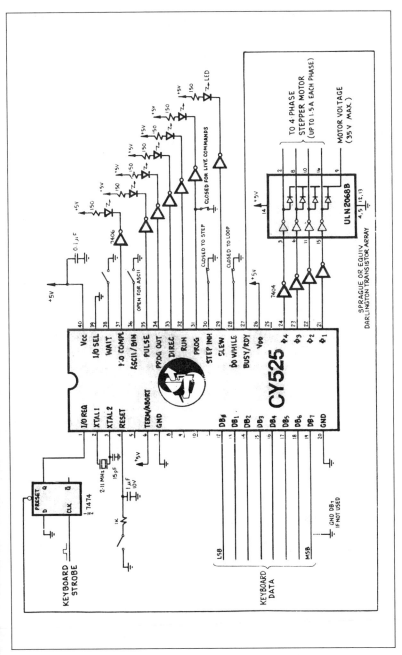

Fig. 4-31. Setup for placing the CY525 in operation. (Courtesy Cybernetics Micro Systems, Inc.)

CHAPTER 5

A Variety of Useful Applications

Ensuing are some applications of power control that should especially appeal to the experimentally inclined and to the natural innovator. These are not blue-sky projects. They have been breadboarded, investigated, and proven technically feasible. For a number of reasons, however, some have not yet attained widespread recognition as everyday techniques for controlling power. Some will probably become more popular as problems of packaging, pricing, production yield, and publicizing are resolved. In other instances, the promise is obviously there, but there remain technical drawbacks to overcome. Also, a strong factor in slowing acceptance of the newer techniques is best identified as psychological—there seems to be a natural reluctance to discard prevailing practices in favor of some new way of doing things. This holds true notwithstanding the merits of newer methods.

It is admittedly true that some of the off-beat techniques have clearly identifiable bugs in them. But it is also true that an evolutionary process tends to overcome and eliminate defective operation once production and marketing get started. Stereo FM broadcasting was such an example. So was varactor turning of TV sets. So were switching power supplies. Further delaying acceptance of new techniques and devices are the misleading and mistaken predictions of experts in the field who seem to miss the boat in much the same manner that persons well-versed in finance and economics misread stock-market trends. Those who have accumulated some years and retain long memories can recall the predictions of otherwise-reliable technology prophets who didn't foresee much chance for the transistor ever competing with tubes in electronics circuits—well, maybe for toys and low-level applications, but certainly not where respectable power-levels must be controlled!

Mindful of these, and numerous other thumbs-down evaluations of those who should know better, some of the following circuits may, indeed, be diamonds in the rough. The intuitive experimenter, as well as the analytical designer, may be able to incorporate some modification or refinement that will make it compelling for all concerned to welcome new approaches to power control.

Because of their essentially experimental nature, some of these circuits will be presented in simplified form, generally without component values and without much implemental guidance or how-to-build discussions. The basic objective in these instances is to present the unique principles underlying the operation of power-control techniques that have not yet become garden-variety.

Other circuits presented are time-tested and sure-fire. They comprise a potpourri of useful power-control applications that did not qualify for inclusion under the titles or dominant themes covered in previous chapters.

REMOTE SWITCHING OF AC-POWERED LOADS

A commonly encountered control situation involves a lamp, motor, or appliance and a nearby source of AC utility power, but a remotely located point from which it is desired to execute switching control. Such a circumstance is to be found in both industrial and residential environments. The traditional solution via a run of conductors in conduit is often both expensive and inconvenient. In order to properly comply with building codes and insurance stipulations, such work is generally done by an electrical contractor. Together with union labor and often formidable wall, ceiling, or floor surgery, such an installation may prove to be far in excess of the apparent triviality of adding a couple of conductors for extending a switching function. Two ways of possibly circumventing a formal conduit-type installation are shown in Figs. 5-1 and 5-2.

In Fig. 5-1, the interface device is the Motorola MOC3011, an optically coupled triac driver. It contains an LED infrared emitter and a gateless triac device which is optically actuated. This internal triac then controls the gate of the large external triac associated with the load.

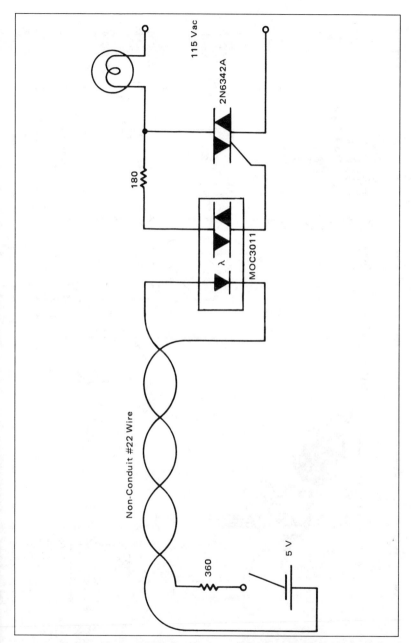

Fig. 5-1. Remote control of a distant AC power-line load by an isolated run of small wire. (Courtesy Motorola Semiconductor Products, Inc.)

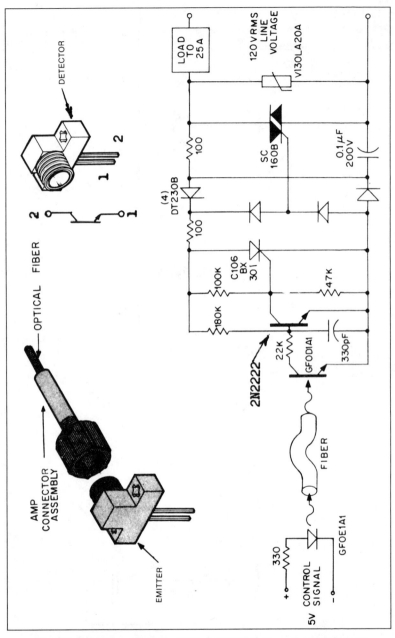

Fig. 5-2. Remote control of a distant AC power line via optical fiber.
(Courtesy General Electric Company)

The MOC3011 can withstand 7500 volts, so it should be well suited to an application of this nature. Nonetheless, it would be good practice to electrically ground one of the #22 wires. This must be done in such a way that there is no ground conflict with the 5-volt DC source. (This probably will not be a battery as depicted.) The optocoupler should not be exposed to the weather. The MOC3011 can supply as much as 100 milliamps for the main triac gate. The 2N6342A triac shown in Fig. 5-1 is a 12-ampere, 200-volt device.

A different implementation of optical coupling for the same problem is shown in Fig. 5-2. Here, the physical separation between the AC-powered load and the control point is made with a run of optical fiber. Because of the optical hardware built into the General Electric optical emitter and optical detector, the installation of this system is much easier than it is when it is necessary to obtain compatible connector elements which, at the same time, must be optically efficient.

Note that the 2N2222 NPN transistor is normally in its conducting state, thereby depriving the SCR of gate current. When light from the fiber impinges on the phototransistor detector, the 2N2222 transistor is turned off; in turn, the SCR receives sufficient gate current to become active. This causes the triac to be turned on and AC is applied to the load. This may not be obvious from the topography of the schematic. However, the four diodes can be redrawn as a recognizable bridge, in which case it will be seen that activation of the SCR also causes current to flow in the gate circuit of the triac. Load and line transients are absorbed by the metal-oxide varistor connected across the triac.

The allowable length of the optical fiber run depends greatly upon the optical attenuation in the fiber. In evaluating manufacturers' specifications on attenuation in optical fiber, make certain that it applies to the 940-nanometer wavelength involved in this application. The experimenter may also wish to investigate the possibility of using a fiber bundle, but must then be prepared to cope with the special connectors or adapters required. Optical fibers operate in various transmission modes and are made of different materials. Efforts to standardize fibers and connectors have only partially succeeded. It is suggested that the circuit of Fig. 5-2 be made operative with a short length of a selected fiber; then experimental data can be obtained for greater lengths. In this way, failure of control can be attributed to optical-fiber attenuation, and not to the electronics of the circuit. It is often discovered that apparently excessive attenuation in the fiber actually stems from alignment problems at the connectors. For this reason, it is unwise to use any splicing techniques when the system is first being checked out for performance.

INTEGRATED HALL-EFFECT DEVICE FOR SWITCHING POWER

Bearing the name of its discoverer, the Hall effect is simple enough. Since the late nineteenth century, it has been known that a voltage can be detected across the side surfaces of a current-carrying conductor if a magnetic field is impressed perpendicular to the path of current flow. This Hall voltage is proportional to current if the magnetic field is held constant. Similarly, the Hall voltage is proportional to magnetic-field strength of the current if held constant. Obviously, a cause and effect relationship exists which could have some interesting and possible useful applications. However, an inhibiting factor to widespread commercial exploitation was the relatively minute voltage exhibited by ordinary metals. Noise, temperature effects, and the need for costly, and often unreliable supporting circuitry were generally cited as valid reasons for avoiding devices based upon the Hall effect.

This has now been dramatically changed, for solid-state Hall elements, together with modern amplifier and logic devices, have brought high performance and reliable Hall-effect devices to the consumer's market. A recent advance in this technology has succeeded in integrating the Hall element and the solid-state electronics on a single chip. It is easy to provide a simple interface circuit so that a large triac can be turned on or off. Practical implementation simply involves a physical situation where a small magnet approaches the IC to produce the desired switching function. One is reminded of similar applications using a reed relay. However, the Hall device experiences no mechanical wear, and can switch at a rate of 100,000 times per second. Obviously, the switching performance is free of the contact bounce that plagues electromagnetic or other mechanical switches.

Typical of the new breed of monolithically integrated Hall-effect devices is the Sprague UGS-3020 Hall-effect digital switch. The block diagram of Fig. 5-3 reveals that in addition to the Hall-effect element (X), this device contains a voltage regulator, an amplifier, a Schmitt-trigger circuit, and two independent output stages. The Schmitt trigger gives the switching function a precise amount of hysteresis so that there is a definite on and off action, thereby preventing oscillation or instability. Hysteresis also gives the switch high immunity to electrical noise and transients. A simple setup for demonstrating the switching action is shown in Fig. 5-4. Magnet speed is not a factor.

This device is intended for service in the real world. It will operate from any DC source over a range of 4.5 to 24 volts, and its output stage

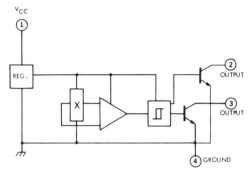

Fig. 5-3. Functional block diagram of the UGS-3020 Hall-effect switch.
(Courtesy Sprague Electric Company)

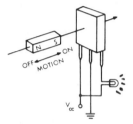

Fig. 5-4. Setup for demonstrating switching behavior in a Hall-effect sensor. (Courtesy Sprague Electric Company)

will sink 25 milliamps. The constant-amplitude output is compatible with all logic families. A small permanent magnet suffices for the actuating element. The switching points are reasonably constant over a wide temperature range; operation can be had from −40°C to +125°C. Strong magnetic fields are not damaging.

A control scheme for switching AC power by varying the proximity of a small permanent magnet to the Hall-effect sensor is shown in Fig. 5-5. Here, a garden variety optoisolator interfaces the UGS-3020 and the triac. Only one output of the UGS-3020 is used. The triac, an RCA-40669, has 8-ampere continuous-current capability when working from a 60-Hz, 120-volt power line. The rudimentary DC power supply utilizing a 6.3-volt filament transformer is more than adequate for this system; it can, in fact, be used to supply operating DC to several additional similar systems.

Practical insight into the sensitivity of the Hall-effect sensor can be gained from an inspection of Fig. 5-6. The south pole of the small Alnico-V rod magnet must face the brand insignia on the face of the Hall-effect sensor. Other than the indicated translatory motion can be

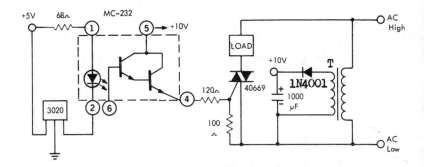

Fig. 5-5. Using a Hall-effect sensor to switch an electrically isolated load. (Courtesy Sprague Electric Company)

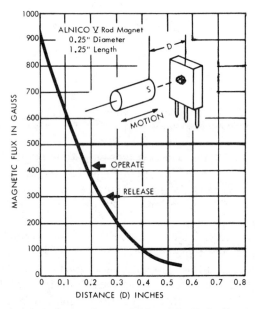

Fig. 5-6. Graph depicting the sensitivity of the Hall-effect sensor.

used. For example, sliding or rotational movements sometimes better fit the mechanical aspects of the system. In any event, the speed at which the magnet moves does not affect switching performance. There is a general tendency for reduction of sensitivity with increasing temperature. This, however, is not pronounced, and in most installations will not need any special attention.

A RUGGEDIZED SOLID-STATE RELAY

For experimental work and commonly encountered laboratory use, a solid-state relay should display the electrical ruggedness usually associated with electromagnetic types. However, the optoisolators generally used in solid-state relays are quite vulnerable to damage. Reverse voltage or excessive forward voltage can easily be catastrophic to the gallium-arsenide LED in these units. Overvoltage, and the resultant overcurrent can deplete the performance of LEDs even if they are not immediately destroyed. All LEDs lose light emissivity during their admittedly long useful life; moreover, useful life can be greatly shortened via abusive operating current. Accordingly, the input circuit of a solid-state relay that is not permanently designed into a circuit or system should have protective circuitry for its input.

Solid-state relays often comprise an optoisolator working into a triac. The triac is subject to increased stress when interrupting inductive loads such as motors, magnets, etc. What happens is that due to the phase displacement between current and voltage, the triac fails to commutate properly at zero-current crossings because the voltage is not also zero. This situation will be tolerated by the triac if the rate of change of voltage is slowed down. This is the function of so-called snubber networks. Ideally, a snubber network should be tailored to a particular inductive load. Practically, however, good snubber action can be attained for a reasonably wide variety of inductive loads.

Fig. 5-7 shows a ruggedized solid-state relay with both input and output protection. Protection against reversed polarity of input voltage is provided by the 1N4002 diode. Protection against excessive forward-

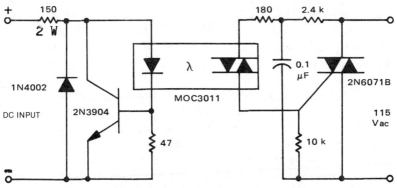

Fig. 5-7. Ruggedized solid-state relay. (Courtesy Motorola Semiconductor Products, Inc.)

polarized input voltage is provided by the transistor connected in the circuit as a shunt voltage regulator; the higher the input voltage, the higher is the LED current, which in turn develops higher forward bias for the transistor, thereby increasing its shunting action. The transistor, by controlling the voltage drop across the 150-ohm series resistor, limits the applied voltage to the LED within safe limits.

The output circuit of this solid-state relay has a snubber network consisting of the 2.4K resistor and the 0.1-μF capacitor. This snubber network protects both the main triac and the drive triac from the effects of inductive loads. It also helps prevent inadvertent triggering of the main triac by power-line transients.

SOLID-STATE RELAYS FOR SWITCHING DC LOADS

Solid-state relays for switching DC circuits advantageously utilize the low saturation voltage drop of bipolar transistors. This drop is much lower than the forward voltage drop across a PN diode carrying the same current. If one were not aware of this, it would seem that the facts should be just otherwise because two PN junctions are involved in the transistor contrasted to the single junction in the diode. One might similarly suppose that the common-base configuration should be used because only one transistor junction would be involved in the load circuit. However, the basic physics of the device causes the saturated voltage drop between emitter and collector to be *less* than between base and collector. Thus, the common-emitter configuration is best for solid-state relays.

Two solid-state relays using common-emitter connected power transistors as contacts are shown in Fig. 5-8. Both make use of an optoisolator to separate input and output circuits. The relay of Fig. 5-8A operates in the normally open mode—without the proper input voltage, no load current can be passed. The relay of Fig. 5-8B simulates normally closed contacts—an appropriate DC input must be applied in order to *interrupt* load current.

In the circuit of Fig. 5-8A there is normally no forward bias applied to the base of the output transistor. If, however, a 5-volt DC source is connected to the input terminals, the phototransistor will be conductive because of the radiation impinging on its photosensitive base. This will cause the normally off PNP transistor to conduct, thereby applying the necessary base bias to the output transistor to drive it into hard conduction. Thus, a voltage applied to the "solenoid" causes the relay "contacts" to close. Note that the phototransistor also has an electrical

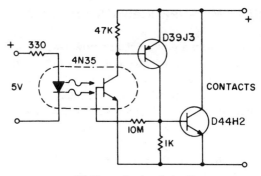

(A) Normally closed circuit.

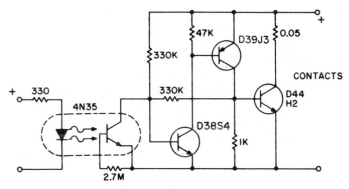

(B) Normally open circuit.

Fig. 5-8. Two 10-ampere, 25-volt DC solid-state relays. (Courtesy General Electric Company)

connection to its base region. In this circuit, the 10-megohm resistor between this base and the base of the output transistor helps speed up contact closure by regenerative action. (If a lower resistance were used, the circuit would behave as a latching relay—the output transistor would remain conductive after removal of the input voltage.)

In the normally closed circuit of Fig. 5-8B, an additional stage has been added to invert the result of activation of the LED emitter. The tiny resistance in the collector lead of the output transistor reflects the fact that maximum dissipation occurs during standby in this relay. This resistance drains off some power dissipation which could otherwise induce thermal runaway, particularly during overload.

In both circuits, the command situation at the DC input terminals can be negated by applying an appropriate bias voltage at the base lead of the phototransistor.

INTERFACING A MICROPROCESSOR WITH AN AC LOAD

The interface between digital logic and power-actuation circuits was discussed in Chapter 1, where the technique of optoelectronic isolation was advocated. Such optical coupling between logic and power sections of a computer system is particularly appropriate, inasmuch as it helps the overall system qualify for UL recognition. Practical examples will be shown for the Motorola TTL compatible microcomputer, the MC3870.

There are two ways to implement the power-control process—load power can be turned on either by a digital logic 0 or a logic 1. Different circuitries are required for the two situations. However, except for this transposition in functional logic, the operating principles of the two circuits are essentially the same. Both, for example, utilize a small transistor as a current booster for energizing the LED within the optoisolator. And, both switch the load power on or off by means of a triac.

The circuit for achieving logic 1 activation of the load is shown in Fig. 5-9. Transistor Q1 is needed because the MC3870 can deliver only about 300 microamps, whereas the LED in the optoisolator needs 10 milliamps. The phototriac is triggered by the optical output from the LED, but otherwise behaves much like the triac trigger device commonly used in triac circuits. Thus, it can be seen how the load receives AC power through the main triac when the MC3870 outputs a logic 1. It should be noted that the firing of the main triac causes a near short to

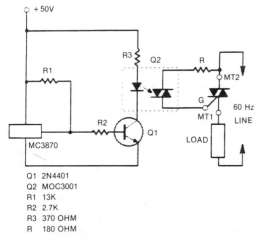

Fig. 5-9. Optical interface circuit for logic 1 activation of load power from the MC3870. (Courtesy Motorola Semiconductor Products, Inc.)

be placed across the phototriac, thereby turning it off; the main triac continues to conduct until commutated by the zero crossing of the AC line current. With the oncoming half cycle, retriggering of both the phototriac and the main triac occurs, providing the logic 1 level continues to be output by the MC3870.

Fig. 5-10 depicts the circuit for achieving logic 0 activation of the load. Instead of the LED current being controlled by an NPN transistor, a PNP transistor is used. Because of this, the LED is energized when the MC3870 outputs a zero logic level. Otherwise, the action within the optoisolator and in the main triac load circuit is exactly the same as in the previously described interface scheme.

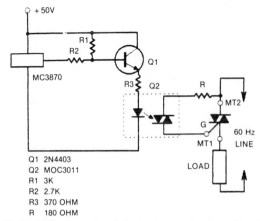

Q1 2N4403
Q2 MOC3011
R1 3K
R2 2.7K
R3 370 OHM
R 180 OHM

Fig. 5-10. Optical interface circuit for logic 0 activation of load power from the MC3870. (Courtesy Motorola Semiconductor Products, Inc.)

Other similar optoisolators may be used but heed must be paid to the LED ratings and to the peak blocking voltage. The optoisolators used in the above described applications are specifically designed as triac drivers. A group of such dedicated optoisolators listed in Table 5-1 enable appropriate values of limiting resistance R to be selected.

A group of suitable load triacs covering a wide current range is depicted in Table 5-2. These are nonsensitive gate types inasmuch as the optically coupled triac drivers can deliver 100 milliamps, and have one ampere peak ratings.

Table 5-1. Specifications for Typical Optically Coupled Triac Drivers (Courtesy Motorola Semiconductor Products, Inc.)

Device Type	Maximum Required LED Trigger Current (mA)	Peak Blocking Voltage	R(Ohms)
MOC3009	30	250	180
MOC3011	15	250	180
MOC3011	10	250	180
MOC3020	30	400	260
MOC3021	15	400	360
MOC3030	30	250	51
MOC3031	15	250	51

Table 5-2. Suitable Load Triacs for Logic-Interface Systems Using Optically Coupled Drivers (Courtesy Motorola Semiconductor Products, Inc.)

Triac	$T_{T(RMS)}$
MAC91,A	0.6
2N6068-75	4.0
2N6342-49	8.0
2N6342A-49A	12
MAC15, A Series	15
MAC223, A Series	25
2N6157-65	30
2N5441-46	40

FLASHERS

Flashers have many uses. They are widely applied as warning devices, visual alarms, and as attention getters. They serve many advertising purposes, and are profusely found in toys, games, and novelty items. They are used in timing and in special illumination techniques, and they sometimes prove useful for instructional purposes. Most flashers either use filamentary lamps or LEDs, although fluorescent and gaseous discharge lamps can be used, too. The switching mechanism, once dependent upon mechanical devices or on various types of tubes, is now almost universally achieved with solid-state devices. The most popular format employs either one or two light-emitting elements; when two are used, they generally go through their on and off cycles alternately. Lenses and reflectors may be used, as the application dictates. Operating power may be derived from self-contained batteries, or from the AC utility line. Common denominator to most flashers is the use of some type of relaxation oscillator or multivibrator circuit; rate, and/or duty cycle may be fixed or adjustable. Sometimes, an audio output may be incorporated with the visual indication, as with certain alarm systems.

The most direct approach to flasher operation is to place visual indicators in the output leads of a conventional RC multivibrator circuit. Such a scheme is shown in Fig. 5-11, where LEDs indicate the conductive states of the two power MOSFETs. Do not use zener-protected MOSFETs for this application. The prescribed cautions must be exercised during handling and soldering in order to protect the gates from static, and from power-line leakage and transients. Note that a constraint on the voltage swing at the gates is imposed by a positive-bias arrangement. There is more than meets the eye in this circuit. The almost infinite gate impedance of the MOSFETs can be exploited to produce very slow sequencing without the use of inordinately large coupling capacitors. However, it will then be found that the insulation quality of certain materials, such as phenolic boards, is inadequate. Indeed, the leakage in some printed-circuit boards may also be found excessive. The experimenter can readily modify this circuit to provide for the use of various filamentary lamps, inasmuch as many power MOSFETs will have adequate current capability for the purpose. The fact that this circuit does not need electrolytic coupling capacitors makes its operation more reliable and more predictable than in an analogous design using bipolar transistors. Suitable power MOSFETs are the VN35AB, VN40AD, VN66AF, and VN88AD.

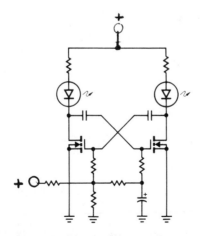

Fig. 5-11. Flasher using power MOSFETs and LEDs in a multivibrator circuit. (Courtesy Siliconix, Inc.)

The 50% duty cycle that would normally accrue from this type of circuit can be altered by making the RC timing constants asymmetrical. Also, not only different-colored LEDs can be used, but entirely different types of lamps may be employed for special applications. For more light output, several LEDs can be paralleled in each drain circuit; however, it is then advisable to use separate current-limiting resistors in the individual LED circuits. And, although not commonly done, series-connected LEDs are also feasible with the advantage that only one limiting resistor is needed.

A BATTERY-OPERATED SCR FLASHER

The flasher shown in Fig. 5-12 is a proven design that has seen considerable use by manufacturers of warning devices. It is well suited for untended battery operation, and its parts count and cost are both low. It is not necessary to use the exact active devices designated in the circuit diagram, and there is considerable latitude in the selection of the battery and lamp. However, before attempting any modification of this basic circuit, a clear understanding of the operating principle should be attained.

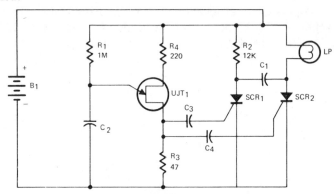

Fig. 5-12. Battery-operated SCR flasher. (Courtesy International Rectifier Corp.)

At first inspection, it might appear that the two SCRs form a multivibrator circuit, with one of them (SCR2) providing the intermittent current for the lamp, and that the flash rate is synchronized to the frequency of the unijunction-transistor oscillator. This, however, is not the operational mode of the circuit. What we have is an SCR that is periodically turned on via a gate pulse from the unijunction-transistor oscillator, then subsequently turned off by a commutating pulse delivered from the *other* SCR. Thus, there is no operational symmetry between the SCRs—one could not, for example, put the lamp in the anode circuit of the other SCR, as would be the case with a multivibrator circuit.

It turns out that SCR1 is the commutating SCR; its purpose is to generate a negative pulse which passes through capacitor C1 and stops conduction in SCR2. A significant aspect of the mode of operation is that SCR1 never remains in its on state much beyond the duration of the narrow gate pulses received from the unijunction-transistor oscillator. This is because resistor R2 does not allow SCR1 sufficient holding current to remain turned on after termination of the gate pulse. This may be construed as the "trick" of this circuit. If it were not for this unique mode, circuit operation would quickly reach a dead end and there would be no repetitive on-off sequencing of the lamp.

Making use of the above information, the best way to visualize circuit operation is to assume the flasher is functioning as intended. Suppose our investigation commences at a moment in time when both SCRs are turned off. A gate pulse from the oscillator then tends to turn both SCRs on. SCR2 turns on and remains on; SCR1 turns on only *momentarily* because of insufficient holding current from R2. It might be thought this momentary turn-on would commutate SCR2, turning it off. It is important to realize, however, that this does *not* occur. The turn-on action of SCR2 is stronger and takes precedence over the tendency of the commutating pulse from SCR1 to turn it off. The *next* gate pulse from the oscillator finds SCR1 in its off state and SCR2 *already* in its on state. Under this condition, the momentary triggering of SCR1 *does* commutate SCR2, turning it off. A subsequent gate pulse causes the cycle to start anew, with SCR2 again being turned on.

If an adjustable flash rate is desired, a variable resistor can be substituted for R1. With the circuit setup as in Fig. 5-12, about 60 flashes per minute will be produced.

A DEDICATED IC SUITABLE FOR FLASHER SYSTEMS

The LM3909 IC provides a novel function—it pulses an LED while operating from a single cell 1.5-volt DC supply. The readily attainable pulse rate is in the vicinity of one flash per second, and this is easily modified upward or downward by appropriate selection of an external timing capacitor. The fact that the forward-voltage drop of LEDs is in the neighborhood of 1.7 volts poses no obstacle, even for run-down flashlight cells. This is because the timing capacitor also participates in a voltage-incrementing action so that the output voltage can be 2 volts, or higher. Although it is true that this basic application does not involve enough power manipulation to qualify for consideration under the intended scope of this book, a quick look at its capabilities is likely to suggest incorporation into systems involving respectable power levels.

Several implementations of this interesting IC in flasher circuits are shown in Fig. 5-13. The experimenter can readily see that the single LED could just as well be the infrared-emitting diode of an optoisolator. In such an instance, drive can be derived for actuating powerful flashers via power transistors or thyristors. Fig. 5-13A shows the basic LED flasher circuit. An alkaline "D" cell should last about two and one half years. A more useful foundation circuit is the variable flash-rate scheme shown in Fig. 5-13B. Here, the flash rate can be varied from 0–20 Hz. The incandescent-bulb flasher of Fig. 5-13C is included to show that higher output current is readily attainable. Thus, there should be no trouble encountered in interfacing this IC with an optoisolator, a transistor, or a thyristor for ultimate control of a moderately high-powered flasher system. By making use of this IC, much experimentation with multivibrators, oscillators, and timing elements can be circumvented. This is worthwhile because it isn't always easy to deal with problems of leakage, inordinately sized elements, low input impedances of active devices, and unreliable startup that are common to circuits operating at flasher rates.

Known tolerance tantalum solid-state electrolytic capacitors are suitable for the timing function. Inexpensive garden-variety electrolytics of the common kind are satisfactory, but be prepared for capacity tolerances as sloppy as −20% to +100%. These may also exhibit too great a temperature dependency in some applications.

INCANDESCENT-LAMP FLASHER USING A THYRISTOR

The incandescent-lamp flasher shown in Fig. 5-14 appears to be an exercise in the use of thyristors, inasmuch as it uses diacs, SCRs, and triacs. However, its practical implementation and its principle of operation are quite straightforward. The heart of the flasher is a multivibrator circuit formed around the two SCRs. The two-terminal diacs are employed in conventional fashion for triggering the gates of the SCRs. Note that the SCRs operate from a DC supply comprised of a half-wave rectifier and a filter capacitor. This means that the SCRs must be forc-

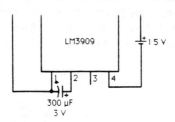

(A) Basic LED flasher circuit. Nominal flash rate, 1 Hz.

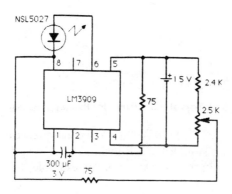

(B) Variable flash-rate circuit. Range, 0–20 Hz.

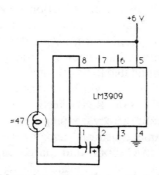

(C) Incandescent-lamp flasher. Flash rate, 1.5 Hz.

Fig. 5-13. Low-powered flashers using a dedicated IC.
(Courtesy National Semiconductor Corp.)

ibly commutated to their off states, i.e., their anode-cathode current paths must experience an interruption or a negative transient long enough to stop conduction. This follows from the fact that a DC-operated SCR cannot be turned off by a gate signal.

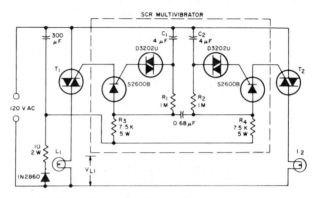

Fig. 5-14. Thyrister multivibrator flashes. (Courtesy RCA)

Commutation of the SCRs occurs as a consequence of transients developed across resistors R3 and R4 when the SCR associated with one of these resistors turns on. Assume, for example, that the left-hand SCR has just turned on; a large negative transient thereby generated across resistor R3 would then be communicated through the 68-μF capacitor to the anode of the right-hand SCR, turning it off.

The turn-on cycles of multivibrator action are governed by the time required for capacitors C1 and C2 to charge to the triggering voltage of their respective diacs; these, in turn, trigger the SCRs which serve as drivers for the lamp-controlling triacs. For example, if the right-hand SCR has just been commutated to its off state, capacitor C2 will charge through resistor R2 until the right-hand diac is triggered. Note that once the right-hand SCR is on, R2 is effectively connected to the negative side of the DC supply, thereby stopping the charging cycle of capacitor C2. At the same time, resistor R1 remains effectively connected to the positive side of the DC supply and charges capacitor C1 until the left-hand diac is triggered. As delineated above, this event abruptly turns off the right-hand SCR. These actions give rise to the alternate conductive state of the two SCRs, and to the corresponding flashing of the two triac-controlled lamps. Timing can be changed by altering the values of R1, R2, C1, or C2. R3 and R4 also affect timing, but these resistances should not be significantly modified because of their commutation function.

When selecting triacs, two ordinarily ignored characteristics of filamentary lamps must be taken into account. First, such lamps have high inrush currents when they are initially turned on because the resistance of the cold filament is much lower than that of the hot filament during normal operation. Inrush current for a 100-watt lamp can be on the order of 10 amperes. Thus, a triac must have in addition to its continuous-current rating, an adequate peak-current rating. Unfortunately, ratings based upon inrush and operating currents only could prove woefully inadequate in the face of abnormal operation resulting from the burn-out of a lamp.

When a lamp burns out, one might suppose that a simple interruption of operating current occurs. However, the parting of the filament is accompanied by vaporization of the tungsten which gives rise to a momentary arc. This is known as *flashover* and it imposes an extremely high current demand; in a 100-watt lamp, the flashover current can be in the vicinity of 100 to 200 amperes. Taking into account the phenomena of inrush current and the shorter duration, but higher-amplitude flashover current, suitable 100-watt lamp-control triacs (T1 and T2) for this flasher are the RCA MAC15-6 types. These triacs are rated for 15-ampere continuous loads and have 150-ampere single-cycle surge capability. (A single cycle at 60 Hz corresponds to 16.7 milliseconds, whereas the flashover duration of a 100-watt lamp is generally less than 4 milliseconds.)

It might appear from the above that triacs for controlling 1000-watt lamps in this flasher would have to have very high surge-current ratings, indeed. Actually, this is not the case; it turns out that the flasher current of 1000-watt lamps remains about the same as for 100-watt lamps. The RCA 2N5808 is a 40-ampere triac with a 300-ampere single-cycle surge rating. It is suitable for control of 1000-watt lamps and its cost is much less than would be extrapolated from the situation involving the 100-watt lamps.

FLASHER USING A PUT OSCILLATOR

The flasher shown in Fig. 5-15 uses a programmable unijunction transistor (PUT) as an oscillator, and is exceptionally easy to build and place into operation. The PUT is an anode-gate thyristor which lends itself particularly well to use in a flasher circuit. This is because a simple PUT relaxation oscillator allows more or less independent manipulation of repetition rate and duty cycle. The configuration of the PUT oscillator is slightly different from that commonly encountered in order to obtain a multivibrator-type waveform instead of sharp pulses. Although the internal operating mechanism is different from an ordinary unijunction transistor (UJT), the two devices develop similar negative-resistance characteristics. Like the UJT it is intended to replace in interest of greater design flexibility, the PUT is a small several-hundred milliwatt device.

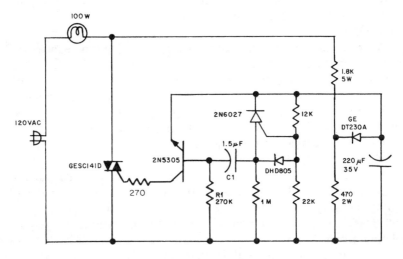

Fig. 5-15. Flasher using a PUT oscillator. (Courtesy General Electric Company)

In the circuit of Fig. 5-15, the triac output stage is interfaced with the PUT oscillator through the 2N5305 trigger circuit. Although not specifically drawn as such, the 2N5305 is actually a small Darlington transistor. The Darlington format imposes minimal loading on the oscillator circuit. When adapting this flasher to a particular situation, the flash rate will be found to depend primarily on capacitor C1. And, the flash duration can be adjusted by selection of resistor R1. Although the lamp triac circuit operates directly from the AC power line, the remainder of the flasher is DC powered. DC is provided by the simple half-wave rectifier circuit comprising the 1.8K and 470-ohm voltage-reducing resistors, the DT230A rectifier diode, and the 220-μF filter capacitor. In implementing this flasher, due consideration must be given to the lack of power-line isolation. (In this respect, it is similar to radios and TV sets.) A desirable safety feature would be a 1:1 power-line isolation transformer. Such a transformer, if not too large, can also provide current-limiting action to protect the triac from lamp flashover and from short circuits; by the same token, the lamp inrush current can be beneficially softened.

STABILIZATION OF LASER DIODE OPTICAL OUTPUT POWER

Many experiments and applications of laser technology are greatly facilitated by using injection laser diodes. These, for practical considerations, are similar to, but involve more sophisticated fabrication than, LED diodes. Indeed, the laser diode operates essentially as an LED at low currents; as current is raised, a threshold is reached at which the optical output changes mode from noncoherent to coherent radiation. In the later mode, the diode is said to be lasing. In the lasing region, current consumption and optical output power are very temperature dependent. In order to protect the diode and to stabilize the optical output, special techniques are necessary. A much-used approach has been to maintain the temperature of the laser diode constant and operate it from a constant-current source. This involves a thermal-feedback system utilizing a temperature sensor, feedback, and a thermoelectric cooler. This is a workable scheme, but a simpler solution to the problem now exists.

Inspection of the stabilization circuit shown in Fig. 5-16 reveals that there are no thermal sensors, thermoelectric coolers, or constant-current sources. Instead, the laser diode has a built-in photodiode that monitors a tiny portion of the radiation. This makes possible an optical-electronic feedback loop in which the current in the laser diode is automatically reduced if the radiation intensity tends to increase from a set level. Conversely, the current is increased when the radiation intensity tends to fall. The overall result is that the optical output is stabilized in the face of a wide range of temperature change. Note that the relationship between current and optical output is linear in the lasing region. Also, it can be seen that without such stabilization, it is possible for a sufficient temperature change to cause nonlasing operation.

In the circuit, the CA3130 op amp performs as a high-gain voltage comparator. It compares a reference voltage at its noninverting terminal, 3, with the voltage derived from the photodiode. The output (pin 6) of the op amp controls the conductivity of a Darlington transistor pair, which, in turn, controls the laser diode current.

A good quality series-pass voltage-regulated supply should be used to provide the 12–15 volts operating DC for this stabilization system.

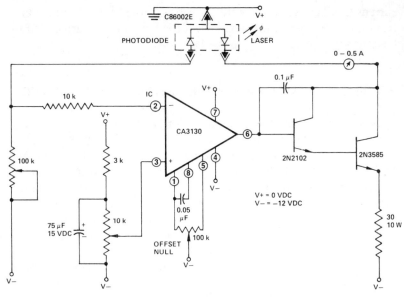

Fig. 5-16. Laser diode system with constant output over a wide ambient-temperature range. (Courtesy RCA)

Switching-type power supplies are not wise choices because laser diodes are susceptible to damage from transients.

WARNING: During operation, this laser diode produces invisible electromagnetic radiation which may be harmful to the human eye. This pertains to rays from reflecting surfaces as well as the direct beam.

AN ELECTRONIC SIREN

The electronic siren shown in Fig. 5-17 is probably best suited for use with toys, inasmuch as the output power level is in the one-third to one-half watt range. However, with subsequent power amplification, very useful siren systems can be implemented for vehicles, boats, and burglar alarms.

In order to simulate the sound characteristics of a mechanical siren, it is necessary to amplitude-modulate a tone by a lower frequency. From practical experience, the basic tone can have a nominal frequency of, say, 1 kHz and the modulating frequency can be in the vicinity of several hertz. In Fig. 5-17, the power op amp portion of the LM389 is con-

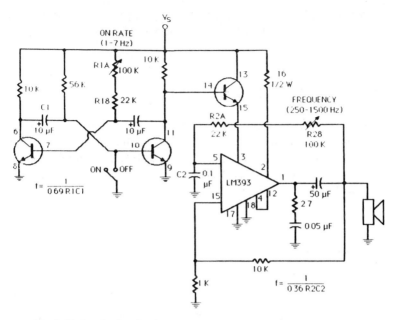

Fig. 5-17. An electronic siren. (Courtesy National Semiconductor Corp.)

nected as a square-wave oscillator to supply the basic tone. Also, two of the on-board NPN transistors are connected in a multivibrator circuit with time constants such that oscillation in the several-hertz range can be obtained. The multivibrator interfaces the op amp oscillator via a third on-board transistor, which serves as the modulator. Modulation is accomplished because this transistor controls the DC current input to the op amp.

Both, the basic-tone frequency, and the modulation rate are manually controllable by variable resistances. A closer simulation to a mechanical siren would probably be obtained from the use of a sine-wave oscillator in place of the multivibrator. Also, an approach in this direction—gradual, rather than abrupt interruption of the basic tone—can be made by experimenting with RC networks in the base circuit of the modulating transistor. The circuit, as it stands, accomplishes the intended function of a siren, which is to capture the attention of the hearer.

Other op amps and other transistors can be used in carrying out the basic design of this circuit. The LM389 is, however, particularly suitable for such an application because of its low quiscent current drain, about 8 milliamps when operating from a 6-volt supply. Optimum power output will result from using a 16-ohm speaker and a 12-volt supply. A factor worthy of consideration in an electronic siren is the acoustic directivity of the speaker.

POWER OP AMP PIEZOELECTRIC BUZZER ALARM SYSTEM

Piezoelectric sound transducers are very efficient converters of electrical energy to acoustic energy. This is particularly true of those intended for use as alarms because they then operate at their self-resonant frequency. A practical approach to implementation of an alarm system utilizing the piezoelectric transducer is to place this device in the positive-feedback loop of an oscillator. If the active device of the oscillator has some power capability, all the elements of a powerful alarm system are then in place.

Shown in Fig. 5-18 is a piezoelectric alarm system configured around a programmable power op amp. The programmable feature enables it to be readily turned on or off from a logic source, if so desired. This op amp has a load-current capability of a quarter ampere. Although this capability is not directly made use of in this oscillator, such a rating puts it in a different league from the general run of op amps. Indeed, at higher DC operating voltages (the absolute maximum power-supply

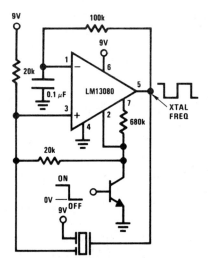

Fig. 5-18. Piezoelectric alarm system. (Courtesy National Semiconductor Corp.)

voltage for the LM13080 is 15 volts), some transducers can be destroyed in this circuit. The intended transducer is the Gulton 101FB. Others can be experimented with, inasmuch as it has become standard practice to make these transducers with three terminals. The terminals are often designated drive, feedback, and ground. In this circuit, the ground terminal connects to the plus side of the DC supply. From an AC standpoint, this does not materially alter its mode of operation. The Archer Cat. No. 273-064 piezo buzzer, which operates at 6.5 kHz is a good element for this circuit (sold in Radio Shack stores).

Any of the common small-signal NPN transistors, such as the 2N2222, can be used as an electronic switch for enabling and inhibiting the alarm. This transistor controls the base bias in the input and output stages of the op amp. As a consequence of this type of control, DC power consumption is very low during inhibit. For best results, the output waveform at terminal 5 should show a 50% duty cycle. The duty cycle can be varied by experimenting with the 100K resistor, or possibly the 0.1-μF capacitor.

Interesting acoustic experiments can be conducted with this circuit. Piezoelectric elements operating at ultrasonic frequencies are also available. These higher frequencies have been used for animal control and insect repulsion. The output of these transducers is quite directional and often can be made even more so with small horns. Between 80 and 90 dB sound pressure level is readily forthcoming. It may be possible to use two-terminal piezoelectric transducers by merely neglecting the AC ground connection. Such simple transducers can be made by simply forming a sandwich of two metal plates and a strip of appropriate barium-titanate material. This will be less efficient than a commercial buzzer, but facilitates experimentation in ultrasonics.

POWER MOSFETS IN AM TRANSMITTERS

The RF power tubes in AM broadcasting stations have always posed reliability problems. They have limited life spans and, at best, slowly die in service from depleted emission. Besides, they are bulky, expensive, difficult to cool, and require high-voltage power supplies. Solid-state design, until recently, has been unthinkable because of the multikilowatt power levels involved. However, practical designs have been demonstrated using large numbers of bipolar power transistors. A drawback to this scheme is that it requires about sixty of these power transistors per kilowatt. Inasmuch as power transistors are physically small, it is feasible to utilize large numbers of them; the difficulty is that bipolar power transistors are not easy to parallel. Problems of current hogging and thermal runaway can be alleviated by the use of ballast resistors, but this seriously impacts operating efficiency and subtracts from the already meager power per unit that is available.

Contrarily, power MOSFETs are readily paralleled in large numbers, and do not normally exhibit thermal runaway characteristics. At the frequencies of the AM RF spectrum, these devices are just loafing along, whereas bipolar transistors incorporate undesirable parameter and ratings trade-offs. Moreover, banks of power MOSFETs remain easy to drive, whereas bipolar types present awkwardly low input impedances when a large number of them are used. And, they do not need energy-wasting decoupling networks to prevent low-frequency oscillation—MOSFET gain is flat from DC to hundreds of megahertz. But, best of all, it only takes eight power MOSFETs to develop a kilowatt of output power.

A simplified circuit of the way in which power MOSFETs may be used to displace tubes in AM transmitters is shown in Fig. 5-19. Note that a bridge-configured RF power amplifier is provided with its operating DC by a switching power supply. By superimposing the audio signal on the error voltage, the output of the switching supply is audio-modulated about its mean voltage level. This, of course, directly modulates the RF power amplifier. An interesting aspect of this scheme is that the IRF350 power MOSFETs suggested by International Rectifier Corp. are not packaged as RF types. The implication is that the fre-

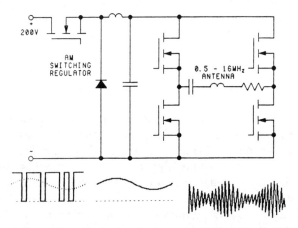

Fig. 5-19. Simplified circuit showing the principle of a power MOSFET AM transmitter. (Courtesy International Rectifier Corp.)

quency capability of these devices is so good that no special techniques are necessary at AM broadcast frequencies.

The IRF350 carries maximum ratings of 400 volts and 15 amperes. The allowable power dissipation at 25°C is 150 watts. Here again, advantages in practical implementation are realized over bipolar transistors, namely that operating impedances tend to be more "tubelike." Finally, the absence of the secondary-breakdown phenomenon inherent in bipolar transistors, frees the power MOSFETs from catastrophic breakdown during modulation peaks.

MODULATION SCHEME FOR POWER MOSFET RF OUTPUT AMPLIFIERS

Another unique method of modulating power MOSFETs is shown in Fig. 5-20. Probably more experimental in nature than the high-level modulation scheme used in Fig. 5-19, this approach is analogous to the grid or efficiency modulation formats long used in tube transmitters. A profound difference from the tube designs is that modulation is accom-

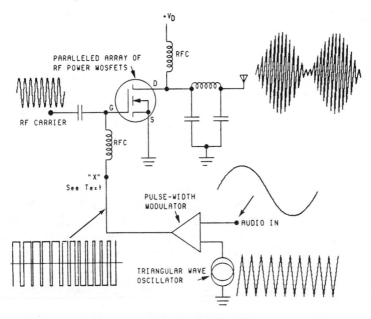

Fig. 5-20. Modulation technique for AM transmitters using MOSFET output stage.

plished via a PWM wave. Note that a switching power supply is not needed in this case; rather, the pulse-width modulated wave is impressed, along with the incoming RF drive, at the gate of the power MOSFET. (Actually, a number of power MOSFETs may be paralleled in order to attain a higher power output.)

A basic requirement of this scheme is that the triangular waves must have a frequency at least several times higher than the cutoff frequency of the low-pass filter in the output circuit of the amplifier. This is so in order to enable the filter to effectively prevent the PWM interruptions from appearing in the envelope of the amplitude-modulated waveform. From a practical standpoint, the low-pass filter must first satisfy the tank circuit and impedance-matching needs of the RF amplifier. Then, a sufficiently high level PWM wave is impressed at the gate to produce a clean modulation envelope at the output. With low-frequency transmitters, this can be readily accomplished, but it becomes increasingly more difficult to do so beyond transmitter frequencies of several megahertz, or so. It then becomes more difficult to produce the requisite high-frequency PWM format, and contradictions in the design of the low-pass filter tend to become more aggravated.

A modification to the arrangement shown in Fig. 5-20 can resolve difficulties such as those described. By inserting, at point "X," a separate low-pass filter with a cutoff frequency just above the audio spectrum of interest, the gate will be impressed with an *audio* signal rather than the chopped PWM wave. The situation will then simulate the grid-modulation technique used in tube transmitters. The output low-pass filter, or pi network, in the drain circuit of the power MOSFET can then be optimized entirely for maximum performance as a tank circuit and impedance matcher. This also relaxes the high-frequency requirements of the triangular wave; it still must be much higher than the highest audio-frequency, but no longer bears any relationship to the transmitter carrier frequency.

Here, again, power MOSFETs can be readily paralleled; this is particularly beneficial in low-level, or efficiency modulation, where RF output power is about one fourth that attainable with high-level modulation. (For power MOSFETs, high-level modulation results from modulating the drain circuit, as is done in Fig. 5-19).

VOLTAGE-REGULATOR IC AS AN AMPLITUDE MODULATOR

The three-terminal IC voltage regulator is sometimes found in other than its intended application as a simple, compact regulated DC supply.

This makes practical sense considering this IC's sophisticated amplifying circuits, stable reference voltage, power-handling capability, thermal protection, and all at low cost. Lambda Semiconductors suggests the unique amplitude-modulation scheme shown in Fig. 5-21. This, or a similar arrangement could conceivably be used to provide AM (amplitude modulation) for a transmitter. (AM continues to be widely used in VHF aircraft communications, and in other services.) Unity or relatively low voltage amplification is obtained, but current and power gain are very high. Frequency response is no problem inasmuch as these ICs can be expected to display flat response to something on the order of 100 kHz, or so.

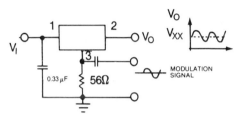

Fig. 5-21. Using a three-terminal voltage regulator as an AM modulator.
(Courtesy Lambda Semiconductors)

Best results will be forthcoming with adjustable voltage regulators, but many fixed types can be anticipated to also work well when used in this fashion. Lambda Semiconductors seems to feel this operational mode should be restricted to ICs with maximum output-current ratings of one ampere. The experimentally inclined will, no doubt, be able to devise ways of safely scaling-up current, if needed. It may well be that paralleling is more easily accomplished in this mode than it is in conventional voltage-regulator applications. And, liberal heat sinking may also prove profitable. In any event, this alternate use for the voltage-regulator IC is an intriguing one; comparing it to conventional modulator designs, it appears likely that a reduction in parts count, cost, and developmental effort can be achieved.

In many three-terminal IC voltage regulators, terminal 1 (in Fig. 5-21) is the input terminal, and terminal 2 is the output terminal. Terminal 3 is generally designated as *ground* in fixed-voltage types, but as *adjust* in adjustable types. The coupling capacitor for the audio signal should be about 1000 μF. Judging from transmitter designs using all discrete devices, it is likely that deep modulation of a class-C RF power amplifier can be accomplished without need of a modulation transformer. (In order to increase modulation percentage, it may be necessary, as is standard practice, to apply some modulation to the driver along with the final amplifier.) In addition to RF transmitters, other applications in instrumentation and in hobbyist activities will, no doubt, suggest themselves. Conceptually, the direct modulation of power-supply voltage is a more elegant design technique than the traditional method of placing a separate modulator in series with the DC source.

POWER MOSFET IN HORIZONTAL-SWEEP CIRCUITS

In the television-servicing industry it has become common knowledge that catastrophic failures are very likely to involve the horizontal output stage. Both power transistors and thyristors have been used in solid-state TV circuitry and both have produced their share of failures. Such failure often destroys other devices and components so that it is not always entirely clear which was the cart, and which the horse. Nonetheless, enough experience has been gained to suggest that reliability would be enhanced if the vulnerability to failure of the horizontal output stage could be reduced. This applies not only to home TV sets, but to other cathode-ray tube display equipment as well.

Many reasons could be advocated to account for this breakdown phenomenon. Obviously the margin of safety designed into the set does not always encompass the simultaneous occurrence of worst-possible operating conditions. Transients, elevated temperatures, thermal cycling, aging, and the effects of both low- and high-line voltage all contribute to reduction of mean time between failure, but the combination is not easy to quantify. The situation is frustrating to the public, and to professionals as well, because (rightly or wrongly) it was widely thought that solid-state equipment would last forever. Moreover, when a solid-state horizontal output device self-destructs, the repair is far more involved, and is costlier than was the simple replacement of a horizontal output tube that had gradually become gassy or lost emission.

Although transistors and thyristors with higher ratings would, no doubt, help matters, this is easier said than done, especially when cost is considered. A possible solution, however, is the use of the power MOSFET. Not only is excellent high-speed performance easy to attain, but considerable circuit simplification results. These devices are easy to drive; the usual power-driver stage and transformer are no longer needed. Best of all, the nature of the SOA curve of these devices is such that failure is less likely. The forgiving nature of the power MOSFET stems partially from the simple silicon output element which involves no PN junction. Then, too, the temperature coefficients are opposite those of PN junction devices which tend to go into thermal runaway if overloaded. The power MOSFET, conversely, tends to *limit* its current consumption. The reason designers haven't hitherto used power MOSFETs in horizontal output stages is because devices with the requisite ratings have not been available. This is no longer true and it appears likely that this device will significantly contribute to the reliability of cathode-ray tube equipment.

The power-MOSFET horizontal-sweep circuit shown in Fig. 5-22 illustrates the lack of clutter generally attending bipolar-transistor and SCR circuits. The turn-off time of the power MOSFET is about 150 nanoseconds compared to 3.5 microseconds or so for the bipolar transistors used for such service. This has many favorable ramifications pertaining both to operation and reliability. The reason for the speedy turn-off is that majority carriers only are involved in MOSFET operation; the minority carriers which produce the undesirable charge-storage effects in PN junction devices have negligible involvement in MOSFETs.

Conspicuous by its absence in the simple circuit of Fig. 5-22 is the phase-locked loop often needed in bipolar-transistor horizontal-sweep systems to compensate for timing errors due to charge-storage phenomena. The requirement for such corrective circuitry stems from the fact that such charge-storage varies from transistor to transistor, and is a complicated function of temperature and operating conditions. The ability to dispense with such extra-baggage circuitry further enhances the operational stability and the reliability of the power-MOSFET horizontal-sweep system.

The drive for the power MOSFET is provided by six logic-circuit inverters connected in parallel; these are contained within a single MOS IC. The rationale underlying this drive technique is to provide a speedy discharge of the gate source input capacitance of the power MOSFET at turn-off time. This type of storage is easy enough to deal with. In contrast, the minority-charge storage in the base-emitter region of bipolar transistors cannot be driven out fast enough to yield a shorter turn-off time than the mentioned several microseconds.

An alternative to the MC1391 horizontal-processor IC is the SGS TDA1180; there are others too, but not necessarily with pin compatibility to the MC1391. The common denominator of these dedicated ICs is that they contain a phase detector and a voltage-controlled oscillator. At this writing, none of these have direct-drive capability for a power-MOSFET output stage. Looking at the overall circuit of Fig. 5-22, it is

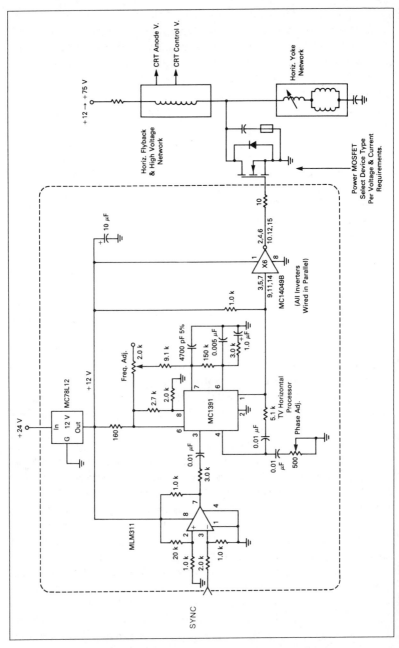

Fig. 5-22. A power-MOSFET horizontal-sweep circuit.
(Courtesy Motorola Semiconductor Products, Inc.)

apparent that an improved horizontal-processor IC could conceivably contain the sync inverter and the power-MOSFET driver. And, judging from what has been accomplished in the power IC field, it appears reasonable to suggest that the 12-volt regulated-DC supply could also be incorporated within the proposed module. What appears to lie immediately ahead are extremely simple horizontal-sweep systems with stable and precise performance, and characterized by very high reliability. Also, much of what has been said also applies to vertical-sweep systems, although electrical stresses tend to be less than in horizontal-sweep systems.

A TWO-TRANSFORMER 100-WATT ULTRASONIC INVERTER

High-frequency inverters are useful as building blocks of regulated power supplies, and as the AC power source for fluorescent lighting, ultrasonic transducers, and specialized welding techniques. Such inverters can be readily implemented as saturable-core oscillators, and often have a single transformer which operates at the output-power level of the inverter. However, with a slightly increased parts count, the inverter can be made to operate more efficiently and more reliably.

The single-transformer saturable-core inverter necessarily sustains high losses from magnetic hysteresis. It also tends to be temperamental with regard to its loading and starting behavior; refusal to start when loaded is an often-encountered phenomenon. Although not always of any consequence, the single-transformer inverter is relatively vulnerable to frequency change as a function of loading. These shortcomings can be minimized via various design techniques, but it is probable that the two-transformer inverter is inherently a better performer. It operates on essentially the same principle as the single-transformer inverter, but a small saturable-core transformer is used in the input circuit. A large transformer is used in the output circuit, but it operates in its linear mode—it is not allowed to saturate.

A practical two-transformer ultrasonic inverter is shown in Fig. 5-23. It delivers 100 watts of square-wave power at an adjustable frequency from 25 to 40 kHz. The nominal output voltage is 140 volts, but this can be readily changed to suit unique needs by altering the secondary turns on the output transformer. The operating efficiency is about 93%. It will be noted that this inverter is operated directly from rectified line voltage—there is no power transformer associated with the rectifier. This considerably reduces weight, bulk, and cost. It also makes overall

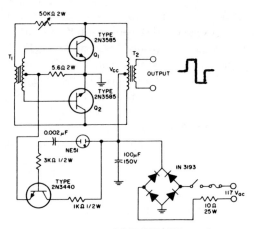

T_1 = Allen Bradley RO-3 (E I 102H 142 A) or equiv.
primary: 160-turn #32 wire;
secondary: each 3-turns #32 wire.

T_2 = Indiana General C2 material (CF216) or equiv.
primary and secondary: 80-turns #28 wire.

Fig. 5-23. A two-transformer, 100-watt ultrasonic inverter.
(Courtesy RCA)

efficiency higher than specifications usually indicate for inverters operating at lower voltages.

Associated with the basic two-transformer inverter is a starting circuit. This auxiliary circuit gives the inverter transistors a jolt of forward bias when the power-line switch is closed. There is no danger of sustained simultaneous conduction of the inverter transistors. This is because of the transient nature of this starting bias. Also, the slightest difference in gain or speed between the two transistors in the inverter quickly causes one of them to take over and initiate normal oscillation.

Frequency is adjustable by the 50K variable resistor in the feedback loop. Although topographically asymmetrical, the circuit is electrically balanced and a good 50% duty-cycle wave is obtained over the entire frequency range.

CAPACITOR-DISCHARGE IGNITION SYSTEM

By the mid-seventies, most automobiles had adopted electronic ignition. Although less standardized than the long enduring breaker-point ignition, the various circuits and formats all yielded beneficial performance features. These included better gas mileage, less susceptibility to fouled plugs, easier starting—particularly in cold weather and with low-battery capability—less tendency to drop out at high engine speeds, and reduced maintenance. There remain, however, many older cars on the road with the more-primitive ignition systems. These cars have not been displaced for a variety of reasons. Nostalgia, economics, and user satisfaction are obvious enough reasons for retaining these cars. Many people view them as higher-quality products than their more-modern descendants. And, certainly, many feel safer in a heavily constructed automobile than in a tiny minicar, fuel mileage notwithstanding. Whatever the reason, most of these older cars can benefit from substitution of an electronic ignition system.

The most convenient type of electronic ignition system for this purpose is one which does *not* dispense with the breaker and condenser within the distributor. This may initially appear as a contradiction inasmuch as one of the declared advantages of electronic ignition is that the breaker points can be eliminated, and with them a lot of headaches and maintenance. However an electronic ignition system that retains the use of the breaker points still drastically reduces maintenance associated with them. This is because the current in the breaker points is so low that there is virtually no arcing, burning, or sparking. Moreover, the clearance between the points is less critical with electronic ignition. The nice thing about retaining the breaker points is that it is then easy at any time to revert back to the original ignition system. Indeed, some of the after-market systems have a switch just for this purpose.

Of the different types of electronic ignitions, the capacitor-discharge type is particularly well suited for upgrading older automobiles. This type of ignition system readily provides a steep wavefront, high-energy pulse that is effective in firing fouled sparkplugs. And it is no trick to design such an electronic ignition system so that, in addition to the breaker-points and condenser, the old ignition coil can also be retained and used. Many of these older cars are not too easy to work on, so the less installation surgery, the better. Some car buffs, to be sure, go all out and install magnetic reluctance or optical-pickup breaker systems, and procure special ignition coils. Most owners will not wish to cope with the extra effort and cost of such refinements inasmuch as the added improvement then becomes marginal in most older cars.

The direct approach to capacitor discharge ignition is to use an SCR to discharge a capacitor through the primary of the ignition coil. There, of course, has to be some means of commutating the SCR inasmuch as DC-operated SCRs ordinarily remain in their conductive state once triggered on. There also has to be a means of quickly charging the storage capacitor to a potential of several hundred volts in order to give the system the requisite energy level. (Incidentally, when ignition breaker points open in the old-style ignition system, the voltage across the primary of the ignition coil is not the nominal 13.5 volts from the battery, but is several hundred volts because of the counter emf from the coil.)

Such an electronic-ignition system generally requires auxiliary circuitry to help stabilize performance over a wide range of conditions—at one extreme, there is the cranking of a cold engine with low battery-voltage; at the other extreme, battery voltage may be at or near its maximum while a fast-running engine demands fast but energetic firing of its plugs. The simplified block diagram of Fig. 5-24 shows the format for a typical capacitor-discharge ignition system using an SCR. (The fact that ignition coils are usually three-terminal autotransformers rather than the shown four-terminal format is of no consequence in demonstrating the basic principles involved.) Between firings, high voltage from the inverter (or more precisely, the converter) charges the storage capacitor; when a plug is to be fired, the SCR is triggered to its on state, thereby dumping the accumulated charge into the primary of the ignition coil. Then the SCR must be quickly turned off in preparation of the next charging cycle. This is straightforward enough, but keep in mind that this cycle of events is repetitive at a very rapid rate in a fast running multicylinder automotive engine.

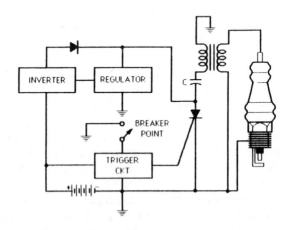

Fig. 5-24. Simplified block diagram of the capacitor-discharge ignition system. (Courtesy RCA)

The circuit of a practical capacitor-discharge ignition system is shown in Fig. 5-25. There are two phases to the construction of this system, the procurement, mounting, and connecting of the components, and the construction of the inverter transformer, T1. The inverter, together with the 1N3195 rectifier diode, becomes a DC–DC converter. Q1 is the switching transistor, and Q2 is its driver in the fairly conventional flyback inverter circuit. Q3 is a control stage; in conjunction with Q4, it provides some of the auxiliary functions alluded to. The first step in building the capacitor-discharge ignition system is to construct the inverter transformer. Table 5-3 lists the parts needed and specifies the

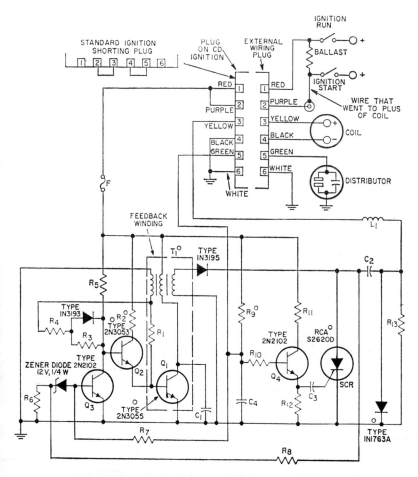

Fig. 5-25. Capacitor-discharge ignition system. (Courtesy RCA)

Table 5-3. Parts List for Capacitor-Discharge Ignition
(Courtesy RCA)

Part	Description
C_1	0.25 µF, 200V
C_2	1 µF, 400V
C_3	1 µF, 25V
C_4	0.25 µF, 25V
F	5 A
L_1 10 µH, 100T of No. 28 wire wound on a 2W resistor (100 ohms or more)	
R_1	1000 ohms
R_2	50 ohms, 5W
R_3	22000 ohms
R_4	1000 ohms
R_5	10000 ohms
R_6	15000 ohms
R_7	8200 ohms
R_8	0.39 megohm
R_9	220 ohms, 1W
R_{10}	1000 ohms
R_{11}	68 ohms
R_{12}	4700 ohms
R_{13}	27000 ohms

All resistors are 1/2W unless otherwise indicated

T_1 = Transformer, wound as follows: A 1/2 in. bobbin and EI stack of grain-oriented silicon steel are used; first, 150 turns of No. 28 wire are wound and labeled start 1 and finish 1 on the winding; second, 50 turns of No. 24 and No. 30 wires are wound bifilar and labeled start 2 and finish 2; third, 150 turns of No. 28 wire are wound and labeled start 3 and finish 3. All windings are wound in the same direction. A total air gap of 70 mil (35 mil spacer) is used. Connections are made as shown in Fig. 5-26.

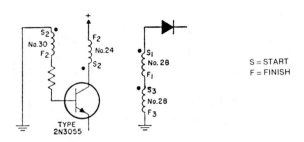

Fig. 5-26. Information for connecting windings on transformer T1.
(Courtesy RCA)

important data for making the transformer. Fig. 5-26 shows how the individual windings are connected. In this type of application, there is some latitude in selection of the dimensions of the E-I segments used in the core of the transformer.

A more-detailed insight into the operation of the circuit is acquired by considering the functions of transistor stages Q3 and Q4. Because of its circuitry position, Q3 has the capability of either shutting down or limiting the amplitude of the flyback pulse delivered to storage capacitor C2. Both of these actions are advantageously used. When the ignition breaker points are open, Q3 completely shuts down the inverter circuit. It does this via sensing resistor R7; when the breaker points are open, the base of Q3 is sufficiently positive to cause heavy conduction in Q3, thereby depriving drive-stage Q2 of operating bias. It follows that the flyback output transistor, Q1, is also inhibited from normal operation. If it were not for this provision, the inverter would have to operate into a short-circuit during those times when plugs were being fired.

Note also that the base circuit of Q3 senses the voltage developed across storage capacitor C2. It does this through R8 and the 12-volt zener diode. Because of this feedback loop, the amplitude of the flyback pulse is regulated; when the voltage in capacitor C2 tends to get too high, Q3 is made sufficiently conductive to partially shut down the inverter by inhibiting its natural regenerative activity. If it were not for this technique, the voltage delivered to the distributor and spark plugs could get high enough to endanger insulation. (Approximately 350 peak volts across C2 corresponds to 40 kilovolts open-circuit at the spark plugs.)

Q4 and associated circuitry provide the trigger signal to the gate of the SCR when the breaker points open. This stage effectively isolates the SCR from stray and transient voltages that might otherwise trigger it. Examples of these false gate signals are those due to breaker-point bounce immediately after closing, residual voltages across the closed breaker points because of imperfect contact, and the inverter frequency riding on the battery supply leads. Because of capacitor C4, the original capacitor across the breaker points can be removed; however, it is desirable to retain it. Not only does it provide additional suppression of spurious voltages, but its retention enables quick reconversion to the original ignition system in an emergency, or during the course of testing and evaluation.

Those items in the schematic diagram which are designated with a small circle should be accorded special heat-removal treatment. Solid-state devices tending to run hot should be heat-sinked. These include the SCR, the 1N1763A commutating diode, and the two inverter transistors, Q1 and Q2. Provide as much convection as possible for transformer T1 and secure power-dissipating resistors to aluminum or copper surfaces with heat-conductive epoxy. Note that the original ballast resistance is automatically shorted out when the new ignition system is plugged in. This is permissible because of the internal electronic regulation.

SIMPLE 12-VOLT BATTERY CHARGER

A four-ampere battery charger is a practical and convenient adjunct for vehicles and boats with twelve-volt batteries. Usually, a battery has been depleted because the lights were left on, or because of excessive use of the starter under difficult starting conditions. In any event, such dead batteries can often be brought up sufficiently for starting purposes within a few hours when charged at a four-ampere rate. At worst, overnight charging is bound to suffice. Although quicker results can be forthcoming at higher charging rates, there is ample evidence that this shortens the lifespan of the battery. Also, higher charging rates tend to produce excessive gassing, and this is sometimes dangerous.

A good design approach for such a charger is a basic series-pass DC voltage regulator with a current-limiting circuit to prevent the charging rate from exceeding 4.5 amps. Such a circuit is shown in Fig. 5-27. It

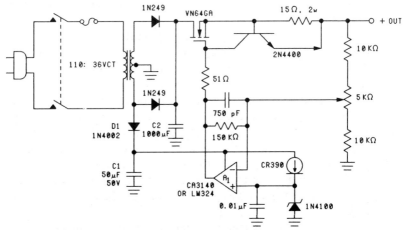

Fig. 5-27. Four-ampere battery charger for 12-volt systems.
(Courtesy Siliconix, Inc.)

uses a power MOSFET as the series-pass element, and an op amp as the error amplifier. This works out nicely because of the meager drive requirement of the MOSFET—if a bipolar transistor were used, at least one additional driver stage would be required. Current limiting occurs when the 2N4400 NPN transistor becomes forward-biased from passage of the charging current through the 0.15-ohm sampling resistor; the 2N4400 then conducts, thereby depriving the power MOSFET of any additional drive voltage.

The error amplifier compares the zener-developed reference voltage with an adjustable sample of the output voltage, and varies the drive to the power MOSFET in such a way as to null the error voltage and maintain the output voltage constant. Note that the zener diode is supplied by a constant-current diode, rather than by the usually used resistor. This enhances the stability of the reference voltage, and also simplifies certain design problems.

Even though the primary of the transformer circuit is fused, and output current is limited to 4.5 amperes, it would be wise to equip the MOSFET with a liberal heat sink. This will protect it from excessive temperature rise which could result from inadvertent short circuit of the output—a not uncommon occurrence under the unfavorable working conditions often accompanying battery failure. Also, although not shown, a 0-10 ampere ammeter will make the application of the charger more meaningful. This is particularly true because some feeling of charging progress can then be obtained—a good battery will accept less charging current as it approaches the fully charged condition.

BURST MODULATION PROPORTIONAL CONTROL OF A HEATING SYSTEM

Burst modulation, otherwise known as integral-cycle control, is particularly well suited for the control of temperature in heating systems. The inherently slow rate of change in such a power system enables very refined feedback control in terms of the number of power-line cycles the load is permitted to be turned on at any given time. Such control is readily made proportional in the sense that the greater the required temperature rise, the greater the number of cycles of power to the heater. This contrasts to the overshoot-undershoot temperature control commonly provided in households via thermostatic switches.

From the electrical standpoint, burst modulation generates negligible RFI or other electrical noise inasmuch as power is initiated and terminated at zero-voltage crossings. Additionally, such control subjects the control thyristor to considerably less electrical stress, thereby enhancing system reliability. And when used for large-lamp control, burst modulation definitely extends filament life because of reduced inrush current.

The circuit shown in Fig. 5-28 is designed around the Motorola MFC8070, a zero-voltage switch capable of directly triggering large triacs. This IC will either deliver gate-trigger pulses for each half cycle of power-line frequency, or will be in a quiescent state. Which of these two states prevails is dependent upon the relative logic levels at two inputs (pins 2 and 3) of the device. In any event, the trigger pulses occur only at zero-voltage crossings.

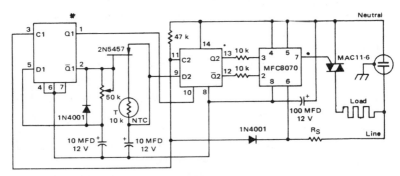

*McMOS Dual D Flip-Flop MC14013L

LINE VOLTAGE	R_S
120 VAC	10 k 2 W
240 VAC	20 k 4 W

Fig. 5-28. Burst-modulation circuit for proportional control of temperature. (Courtesy Motorola Semiconductor Products, Inc.)

The other two function blocks actually comprise a single IC, a dual D flip-flop, the Motorola MC14013L. This IC is made with MOS logic. The two flip-flops are used to initiate and terminate voltage-ramps in two RC circuits and to inform the zero-voltage switch when to turn off the thyristor gate pulses. Significantly, one of the two RC circuits contains a thermistor—its charging rate is therefore dependent upon the temperature sensed by the thermistor. The other RC circuit contains a 50K variable resistor by which its charging time can be manually accordingly, at the next clock pulse, D1, being deprived of this threshold voltage level, makes flip-flop #1 again reverse its state. Thus $\overline{Q1}$ reverts justed. Note that both flip-flops receive line-frequency clock pulses from half-wave rectification via the R_s–1N4001 circuit.

The above has delineated the functions of the important components in this system. To more clearly see how they work together, reference should be made to the circuit and to the waveform diagram of Fig. 5-29.

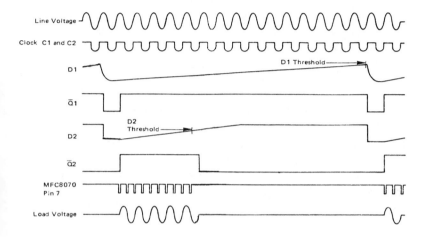

Fig. 5-29. Waveforms for circuit of Fig. 5-28. (Courtesy Motorola Semiconductor Products, Inc.)

Commencing at the left part of the waveform diagram, it will be seen that the initial state of flip-flop #1 is such that $\overline{Q1}$ is high and $\overline{Q2}$ is low. Note that the voltage level at D1 is rising. The first significant event is the attainment of threshold level by the charging voltage at D1; this reverses the logic state of flip-flop #1, making $\overline{Q1}$ low. Referring back to the circuit, it should be appreciated that as long as $\overline{Q1}$ is high, both RC circuits are being charged, that is, their respective 10-μF capacitors are developing voltage ramps. However, when $\overline{Q1}$ goes low, both RC circuits are discharged—one through a 1N4001 diode, and the other through the gate junction of the 2N5457 JFET.

The discharge of the RC circuit associated with flip-flop #2 is sensed at D2 and causes a reversal of logic state in flip-flop #2. Thus, the next significant event is the transition from low to high logic level at $\overline{Q2}$. Because the zero-voltage switch now senses a reversal of logic levels at its input, it is enabled to generate gate trigger pulses for the triac.

A subtle, but very important event in the sequence should now be recognized. When $\overline{Q1}$ became low, it discharged both RC circuits. Ac-to logic high and starts a new charging cycle for the voltage ramps at D1 and D2.

Proceeding to the right on the waveform diagram, the next significant event is the attainment of threshold voltage level at D2, changing the state of flip-flop #2 so that $\overline{Q2}$ is again at logic low. This event restores the zero-voltage switch to its inhibit state, and power to the load is turned off. After the elapse of more time, the voltage at D1 attains its threshold level and the initial events then repeat. Thus, the extreme right-hand portion of the waveform diagram resembles the extreme left-hand portion where the analysis commenced.

Note that the *duration* of applied power to the heater and the length of time *between* such applications are both dependent upon charging rates. A low ambient-temperature (cold thermistor) slows the charging rate at D2, thereby delaying termination of heater power. And increasing the 50K variable resistor (corresponding to a lower temperature requirement) slows the charging rate at D1, thereby delaying initiation of heater power. In a well-engineered heating system, the time between wavetrains of power would not be too long and a person would experience smaller increments of temperature change than is usually the case with the commonly used thermostatic-switch type of control.

A PWM AUDIO AMPLIFIER

Not too long ago, hi-fi enthusiasts became understandably excited over a unique concept of audio power amplification. The technique promised very high efficiency, a clean solution to thermal problems, and worthwhile reduction in the space and weight required per watt. Moreover, it was projected that before long, cost per watt would also be lower than with conventional power amplification. Although performance was not as good as with conventional methods, it was demonstrably acceptable, and there appeared no theoretical reason why it could not be expected to be improved with time. Many prototypes, and even a limited number of commercial models, convincingly showed the scheme to be practically feasible—or almost so. As might be expected with new models in almost any technology, there were, admittedly, some bugs to be contended with. For a number of reasons, and not all of them technical, the off-beat audio system did not survive the setbacks encountered in the introductory models. The basic idea has remained in the contemplations of engineers, and experimenters feel that practical implementation will again see the light of day.

The technique referred to is class-D power amplification, otherwise known as the pulse-width modulation amplifier. Although it did not successfully invade the domain of hi-fi and stereo in its first attempt, this technique has enjoyed widespread use in regulated power supplies, and to a lesser extent in servo systems involving the control of motors. A great deal has been learned, and many improved and dedicated solid-state devices have become available to implement such systems. It is not farfetched to suppose that renewed investigation of the possibilities of this approach to audio power may produce fruitful results.

The principle of the PWM audio amplifier is surprisingly simple. As seen in Fig. 5-30, a pulse-width modulated format is generated when an overdriven summing amplifier is simultaneously impressed with the audio signal and a high-frequency triangular wave, known as the carrier. The class-D amplifiers are then turned on and off by the pulse-width modulated wavetrain. Here is where efficiency enters the picture; because the class-D amplifiers are either on or off but never in an in-between state, their theoretical efficiency is 100%. (This, of course, is based on the ideal premise that their on resistance is zero, that they make instantaneous switching transitions, and that their off resistance is infinite.) The output of the class-D amplifiers is passed through an integrator in the form of a low-pass filter with a cutoff frequency above the audio range, but much lower than the carrier frequency. What emerges is the restored audio wave, stripped of the interfering carrier frequency.

In Fig. 5-30, The Schmitt trigger reproduces the PWM wave, but with steeper edges than when it emerges from the summing amplifier. It does this by delivering an essentially regenerated version of its input. The basic idea is that more rapid switching transitions make the class-D amplifier operate more efficiently—there is less power dissipated during rise and fall times. A driver stage is included to ascertain that the power MOSFETs in the class-D amplifier are turned on hard during their on periods. The low-pass filter which is inserted between the class-D amplifier and the speaker load has a cutoff frequency higher than the audio spectrum, but much lower than the carrier frequency (the triangular wave). As mentioned, this filter performs integration, or *demodulation* of the PPM wave. The wave finally impressed across the speaker is a near-replica of the original audio-input signal. Also shown in this diagram is an input amplifier for the audio signal. Whether this is needed for the sake of gain will depend upon the circumstances of an individual system, but it does provide an electrically convenient point to return negative feedback.

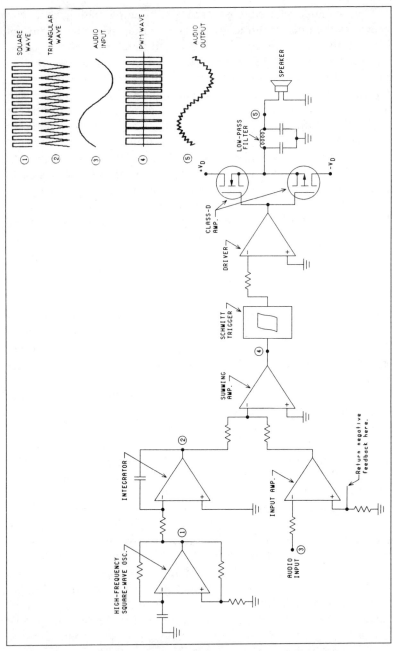

Fig. 5-30. Class-D audio amplification by pulse-width modulation.

It is only fair to comment on some of the difficulties experienced with previous attempts to exploit the features of this unique approach to audio amplification. The parts count—primarily the number of ICs needed—while not excessive, could be dramatically reduced in a modern design by using one of the many dedicated IC control systems for PPM switching power supplies. These ICs incorporate all of the needed circuit functions prior to the class-D output amplifier itself. Their claimed PWM frequency capability is better than 100 kHz, but accompanying charts show that this might be extended to several hundred kilohertz. (The error signal of a switching regulator would correspond to the audio input signal of the PPM audio amplifier under discussion.) That these ICs are applicable can be seen from the block diagram of a typical one shown in Fig. 5-31. They generally contain 50-100 equivalent transistors. It appears that they could greatly simplify a PWM amplifier.

It is necessary that a very high-frequency triangular wave (carrier) be used. This contributes to the ability of a simple low-pass filter to "strain out" the PWM wave so that the recovered audio wave is nearly free of high-frequency contamination. Previously, complementary pairs of power MOSFETs left something to be desired in the way of margins of safety. Now, however, both N-channel and P-channel types are electrically rugged and are available at consumer-oriented prices. These are used instead of bipolar transistors because it is easier to attain the switching speed needed to accommodate the several hundred kilohertz carrier switching rate.

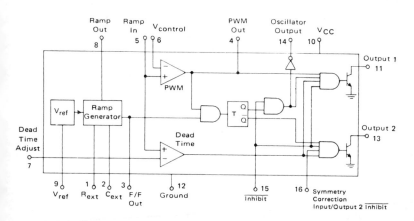

Fig. 5-31. Functional block diagram of MC3420 regulator control IC.
(Courtesy Motorola Semiconductor Products, Inc.)

USEFUL APPLICATIONS / 241

Another reason the carrier frequency must be high is that the effective negative feedback decreases with rising audio frequency to the extent that it is difficult to obtain reasonable feedback at higher audio frequencies. Negative feedback is needed for analogous reasons to its need in conventional amplifiers—the wave processing departs somewhat from ideal performance, so negative feedback is needed to cancel the inadvertent distortion.

Other than a suggested connection point, no feedback provision is depicted in Fig. 5-30. This is because it is felt that this remains a tricky, if not unsolved, problem. It is here that experimentation and innovation can be expected to pay high dividends. A feedback network has the same demands made on it as in conventional amplifiers, but they are more difficult to comply with in this system. Phase and amplitude of the fed-back signal are not so easy to control. If hi-fi performance is not the objective, feedback can be dispensed with.

There are other problems, too. EMI and RFI must be dealt with in a systematic manner. Among other things, this implies good RF-shielding practices. Fortunately, much has been learned about this problem from switching power supplies. From the manufacturer's point of view, it is felt that the necessary shielding, bypassing, and decoupling techniques tend to, at least partially, offset cost advantages in other areas, such as heat removal.

Although this technique may remain controversial or latent for application to hi-fi audio for the consumer's market, its advantages are often exploited where rigorous distortion specifications are not in effect.

Index

AC impedance in regulated power supplies, 6-7
AC motors. *See* Induction motors; Synchronous motors
AC power
  interfacing of, with microprocessors, 204-206
  switching of, 194-200
AC tachometer function, 161-166
Adjustable voltage regulators, three-terminal, 44-45
Alarm system, 218-219
Alternating-current motors. *See* Induction motors; Synchronous motors
AM. *See* Amplitude modulation
Amateur radio amplifiers
  2-meter band, 25-watt, 94-97
  HF band SSB, 97-100
Amplifiers
  amateur radio, 94-100
  audio. *See* Audio amplifiers
  channel, 12-watt, 71-73
  class D power, 238-242
  mobile, 68-71, 88-94
  motor-drive, 84-86
  stereo. *See* Stereo amplifiers
Amplitude modulation
  for RF output amplifiers, 221-222
  transmitters, 219-221
  by voltage-regulator IC, 222-224
Analog oscillator, and three-phase power format, 25-26
Antiparallel mode, SCR, 174-175
Apparent power in polyphase power systems, 22
ASCII keyboard control of stepping motors, 185-187
Audio amplifiers, 62-64
  20-watt, 73-75
  bridge, 66-68
  high-frequency considerations in, 53-54
  with IC-driven output stage, 60-watt, 78-80
  mobile, 68-71
  pulse-width modulation, 238-242
Audio power ICs, 54-56, 68-71
Automobile audio systems. *See* Mobile amplifiers

Automobiles, capacitor-discharge ignition systems for, 228-232
Battery charger, 12-volt, 232-235
Bidirectional servo drive system, 148-151
BIMOS
  bridge inverter, 135-139
  power switch, 39-41
Bipolar transistors
  Darlington. *See* Darlington transistors
  disadvantages of, 32-33, 35-36
  emitter-collector voltage drop of, 202
  germanium, voltage drop of, 112-113
  and motor control, 171-172
  paralleling of, 36, 220
  RF power limitations of, 31-35, 39-40
  in solid-state relay, 202-203
  in stereo amplifiers, 16-18
  in voltage regulators, 105-107
Board layout and RF construction, 96-97
Bridge, synchronous rectifier, 113-116
Bridge amplifiers, 66-68
Bridge inverter, BIMOS, 135-139
Burst modulation, 235-238
Bypassing, high-frequency, 43-44
CA3020 IC, 84-86
CA810Q/TBA810S audio power IC, 69-71
Capacitance
  gate-input, and RF Power MOSFETs, 37
  multiplier, 103-105
  package, of bipolar transistors, 32
Capacitor-discharge ignition systems, 228-232
Capacitors
  elimination of, in switching power supplies, 102
  filter, and amplifier performance, 18-19
  high frequency bypassing by, 43-44
Centrifugal switches, elimination of, 168-169
Channel amplifier, complementary-symmetry, 12-watt, 71-73
Charge pump, 108
Chopper power supplies. *See* Switching power supplies
Class D power amplifier, 238-242
Clock
  in CY525 IC, 186

Clock — cont.
  in digital logic control circuits, 29
CMOS logic ICs, and motor control, 143-145.
    See also Digital logic circuits
Coaxial packaging, 35
Commands for CY525 IC, 186, 189
Complementary-symmetry amplifiers
  channel, 12-watt, 71-73
  output stage, 16-18
  stereo, 40-watt, 75-77
Conduction, heat, 48
Constant-current diodes, 107-108
Construction considerations
  printed circuit board leakage, 207
  RF circuits, 96-97
  with three-terminal regulators, 49, 52
Convection, heat, 48
Converters
  forward, 123
  step-down, 128-129
Copper foil as heat sink, 47-48
Copper oxide rectifiers, disadvantages of, 117
Counter emf and motor control, 143, 159-160
Coupling, interstage, using optoisolators, 56-58
Critical lead resistances, 49, 52
Current
  in polyphase power systems, 20-21
  regulation, 7, 103-104, 107-108
Current foldback, 119-123
Current-fed inverter in motor control, 182, 184
CY525 stepping-motor controller IC, 185-191
Darlington transistors
  in linear power supplies, 9, 102-105
  and motor control, 172
  in stereo amplifiers, 16-18, 75-77
DC load switching, solid-state relay for, 202-203
DC motors, control of, 14
  direction
    using digital logic, 173-174
    using steering diodes, 148
  speed
    using digital logic, 143-145, 163-164, 166
    using op amps, 155-157
    using optocouplers, 161-166
    using phase-locked loops, 161-166
    using PWM module, 142-143, 145-148
Dead band in servo systems, 148
Delta-connected loads in polyphase power systems, 21, 24
Diacs as triggering device, 170
Diac-triac combinations, triggering characteristics of, 148-149, 152
Digital logic circuits
  motor control using, 143-145, 163-164, 166, 173-174
  optical isolation of, 166-168
  power interface for, 38-39
  quadrature power format generated by, 30-32
  switching power supply for, 129-131
  three-phase power format generated by, 28-30

Diodes
  fast-recovery, in motor control, 147
  germanium, 117
  injection laser, 215-216
  junction, characteristics of, 116
  parasitic, in MOSFETs, 115-116, 119
  steering, and motor control, 148
  three-terminal regulators protected by, 45-46
  two-terminal constant-current, 107-108
Direct-current motors. *See* DC motors
Discrete components
  advantages of, 3
  disadvantages of, 10
Disks, slotted, and optocouplers, 161-163
Electronic filters, 103, 105
  and amplifier performance, 18-19
Electronic ignition systems, 228-232
Electronic inverter and motor control, 178
Electronic switch, motor starting, 168-169
Electronically controlled rectifier, 116-120
EMI (electromagnetic interference), 145, 182, 242
Emissivity of surfaces, table of, 48
Fast-recovery diodes, in motor control, 147
Fibers, optical, 197
Filamentary lamps, disadvantages of, 38, 161, 213
Filters, electronic. *See* Electronic filters
Flashers, 206-215
Flashover in filamentary lamps, 213
Flip-flops used in three-phase power format, 28-29
FM systems, amplifiers for, 88-94
Foldback, current, 119-123
Formats of polyphase power systems, 22-31
Forward converter, 123-124
Four-phase stepping motors, control of, 185-191
Frequency divider, programmable, for motor control, 162
Frequency modulation systems, amplifiers for, 88-94
Frequency-voltage relationship in induction motor control, 179
Full-wave phase motor control, 151-155
Gain, power, effect of inductance on, 32-33
Gate puncture and MOSFETs, 37
Gate turn-off thyristors, 40
Grid modulation, MOSFET simulation of, 222
Ground connections, and three-terminal regulators, 46-47
Ground loops
  and optoisolators, 57
  and output ripple, 49, 52
GTO (gate turn-off) thyristors, 40
Hall-effect devices, power switching with, 198-200
Ham radio. *See* Amateur radio amplifiers
Harmonics in polyphase power systems, 24
HC2020H op amp IC, 86-88
HC2500 op amp IC, 155-156

Heat sinks, 47-48, 129
  table of, 50-51
Heating system control, 235-238
HEXFETs, 40
  switching power supply using, 131-135
HF bands, amateur radio, amplifier for, 97-100
High-frequency bypassing, capacitors for, 43-44
High-level language control of CY525 IC, 185
Horizontal-sweep circuits using power
  MOSFETs, 224-227
Hotspotting, transistor, 34, 36
Hybrid power-boost module, 94-97
IC-driven audio output stage, 60-watt, 78-80
ICs (integrated circuits), 3
  advantages of, 9-10
  audio power, 54-56, 68-71
  manufacturing considerations of, 6
  sockets, integrity of, 61
  thermal limitations of, 4
  voltage regulators. *See* Three-terminal IC
    voltage regulators
Ignition systems, capacitor-discharge, 228-232
Impedance, AC, in regulated power
  supplies, 6-7
Incandescent-lamp flasher, 210-213
Inductance
  of conductors, 89
  lead, in bipolar transistors, 32-34
Induction motors
  characteristics of, 168, 177
  direction control, 28
  electronic starting switch for, 168-169
  speed control, 25, 155-156, 177-184
Injection laser diodes, 215-216
Inrush current of filamentary lamps, 213
Insect repulsion, ultrasonic, 219
Integral-cycle control, 235
Integrated circuits. *See* ICs
Intercom system amplifier, 54-56
Interface
  microprocessor and AC load, 204-206
  optical, 39
  power, and digital logic control circuits,
    38-39
Inverter
  BIMOS bridge, 135-139
  current-fed, in motor control, 182, 184
  electronic, and motor control, 178
  polarity, 108-109
  ultrasonic, 100-watt, 227-228
IRF35X MOSFET family, 37, 220-221
Isolation, power
  using Hall-effect devices, 198-200
  using optical devices, 38-39, 56-58, 166-168
  transformers for, 214-215
Keyboard control of CY525 IC, 185-187
L272 dual power op amp, 172-173
LAS6300 pulse-width modulator IC, 130-131,
  146-147
LAS6380 switching regulator IC, 128-129
Laser diodes, injection, 215-216
Lead inductance in bipolar transistors, 32-33

Leakage in printed-circuit boards, 207
LEDs (light-emitting diodes)
  damage to, 201
  as optical source, 56-58, 161
  voltage drop of, 117
Light-controlled thyristors, 167-168
Linear amplifier, 300-watt, 2-30 MHz, 97-100
Linear power supplies, 7-12, 101-102
  using bipolar transistors, 102-107
  constant-current, 107-108
  with current foldback, 121-123
  electronically controlled rectifier, 116-120
  polarity inverter, 108-109
  synchronous rectification, 111-116
  tracking regulator, 109-111
LM379 op amp IC, 64-65
LM384 audio power IC, 55-56
LM388 audio power IC, 54-55
LM389 op amp IC, 216-218
LM391 amplifier IC, 78-81
LM723 IC regulator, 10-15
LM3909 IC, 209-211
Load switching
  AC, remote, 194-197
  DC, solid-state relays for, 202-203
Load triacs, table of, 206
Lockup, counter, in digital flip-flops, 28
Logic. *See* Digital logic circuits
Logic-actuated switch, 166-168
Low impedance power MOSFET driver, 58-60
Manganin wire, 120
MC1391 horizontal-processor IC, 225-227
MC3420 regulator control IC, 241
MC3870 microprocessor, 204-205
MC14528B multivibrator, 142-145
Metal-oxide varistors, 167
MFC8070 switch, 236
MHW252 RF power amplifier module, 94-96
Microprocessors. *See also* Digital logic circuits
  control of CY525 IC by, 186
  DC motor control by, 174
  interfacing of, with AC load, 204-206
  MC3870, 204-205
  power interface for, 38-39
Mobile amplifiers, 68-71, 88
  10-watt, 175-MHz, 91-92
  20-watt, 470-MHz, 92-94
  25-watt, 90-MHz, 89-91
  70-watt, 50-MHz, 89-90
MOC3011 optically coupled triac driver,
  194-197
Modules, 2
  hybrid power-boost, 94-97
  PWM, DC motor control with, 145-147
MOSFETs
  in amplitude modulation transmitter,
    219-222
  in battery charger, 232-235
  in class-D amplifier, 239-240
  disadvantages of, 37
  in electronically controlled rectifier, 116-120
  in flasher, 207-208

INDEX / 245

MOSFETs — cont.
in horizontal-sweep circuit, 224-227
impedance driver for, 58-60
in linear amplifier, 300-watt, 2-30 MHz, 97-100
and motor control, 142-145, 172
ON resistance of, 119
paralleling of, 145, 220
parasitic diodes in, 115-116, 119
in polarity inverter, 108-109
RF power from, 35-37
in synchronous rectifier bridge, 113-116
voltage drop of, 40
Motor control, 13-14. *See also* DC motors; Induction motors; Synchronous motors; Universal motors
Motor-drive amplifier with 120-volt output, 84-86
MOV (metal-oxide varistor), 167
MTH4ON05 MOSFET, 119
Multiple-output switching power supplies, 129-135
Negative voltages, tracking of, 126-128
Neutralization of RF bipolar transistors, 33
ON resistance of power MOSFETs, 119
Op amps (operational amplifiers), 10-11
HC2020H, 86-88
motor control with, 155-157, 172-174
programmable, in alarm system, 218-219
in three-phase power format, 25-26
OPB837 slotted optical switch, 162-164
Operational half-wave rectifier, 116
Optical coupling, 39, 194-197
Optical fibers, 197
Optically coupled triac drivers, 194, 197
table of specifications for, 206
Optocouplers and optoisolators, 201-202, 205-206
and DC motor control, 161-166
interstage coupling by, 56-58
Oscillator, analog, and three-phase power format, 25-26
Packaging techniques for RF power transistors, 32-35
Parasitic diodes in MOSFETs, 115-116, 119
Phase-control of motors, 13-14, 149-155
Phase-locked loops and DC motor control, 161-166
Phasing considerations in polyphase power systems, 24-25
Photoconductive cells, problems with, 161
Photocouplers. *See* Optocouplers and optoisolators
Photosensitive detectors, 56
Phototriac in optical interface, 39
Piezoelectric buzzer alarm system, 218-219
Points, breaker, and electronic ignition systems, 228-229
Polarity inverter, 108-109
Polyphase power systems
advantages of, 19-20
current in, 20-21

Polyphase power systems — cont.
delta-connected loads, 21, 24
formats of, 22-31
power in, 21-22
voltage in, 20-28, 30
Y-connected loads, 21-22
Power amplification, class D, 238-242
Power factor in polyphase power systems, 21-22
Power gain, effect of inductance on, 32-33
Power interface, and digital logic control circuits, 38-39
Power supplies, 6-13. *See also* Linear power supplies; Switching power supplies
Power switch
BIMOS, 39-41
with Hall-effect device, 198-200
Power-boost module, hybrid, 94-97
Precision resistors in voltage regulators, 106
Printed-circuit boards, leakage in, 207
Programmable frequency divider for motor control, 162
Programmable unijunction transistors in flasher circuit, 213-215
Programming of CY525 IC, 185, 188
Pulse transformer for SCR triggering, 175
Pulse-width modulation
for audio amplifier, 238-242
motor control using, 142-143, 145-147, 179, 181-182
PUTs (programmable unijunction transistor) in flasher circuit, 213-215
PWM. *See* Pulse-width modulation
Quadrature power format, digitally generated, 30-32
Quasi-complementary-symmetry amplifiers
output stage, 16, 18
stereo, 120-watt, 80-84
Radiation, heat, 48
RC bridge network, and three-phase power format, 27-28
Reactance in polyphase power systems, 22
Rectification, synchronous, 111-116
Rectifiers, electronically controlled, 116-120
Regulated power supplies, 6-13. *See also* Linear power supplies; Switching power supplies
Regulation, current, 7, 103-104, 107-108
Regulators, three-terminal voltage. *See* Three-terminal IC voltage regulators
Relays, solid-state, 201-204
Remote switching of AC-powered loads, 194-197
Resistance, critical lead, 49, 52
Resistors, divider, in voltage regulators, 106
RF construction, considerations in, 96-97
RF power, 31-37
RFI (radio frequency interference), 145, 182, 235, 242
Ripple, output
and ground loops, 49, 52
rejection of, by three-terminal regulators, 42

Saturation voltage of MOSFETs, 37
Schematic symbols, 4
Schottky diodes, 117
SCRs((silicon controlled rectifiers)
  antiparallel mode, 174-175
  in flashers, 208-209, 210-213
  in ignition system, 228-232
  motor control using, 13-14, 151-155, 157-161, 174-176
  triacs compared to, 151
Selenium rectifiers, disadvantages of, 117
Series-pass voltage regulators, 103-104, 232-235
Servo drives, 84-86, 148-151
Shunt voltage regulators, 103-107
Silicone grease, heat dissipation by, 48
Single-phase power systems compared to polyphase, 19-20
Single-sideband amplifier, 97-100
Siren, electronic, 216-218
Six-step method for three-phase induction motor control, 179-181, 183
Slotted disks and optical switches, 161-164
Snubber networks, 201-202
Sockets, IC, integrity of, 61
Soldering irons and MOSFETs, 37, 207
Solid tantalum capacitors, high frequency bypassing by, 43-44
Sound IC, television, 60-61
SSB (single sideband) amplifier, 97-100
Stabilization circuit for laser diodes, 215-216
Static electricity and RF Power MOSFETs, 37
Steering diodes in motor control, 148
Step-down converter, 128-129
Stepping motors, four-phase, control of, 185-191
Stereo amplifiers, 16-18
  6-watt per channel, 64-65
  complementary-symmetry, 40-watt, 75-77
  power supplies for, 18-19
  quasi-complementary-symmetry, 120-watt, 80-84
Stripline elements, 33-35, 93-94
Surface emissivity, table of, 48
Switches
  centrifugal, elimination of, 168-169
  logic-actuated, 166-168
Switching, remote, of AC-powered loads, 194-197
Switching power supplies, 101-102
  100-watt, 100-kHz, 123-126
  bridge inverter, 135-139
  using LM723 IC, 13
  with negative voltage tracking, 126-128
  switching frequency of, 102
Synchronous motor control
  direction, 28
  speed, 25, 155-156
Synchronous rectification, 111-116
Synchronous speed, single-phase induction motor, 168
Tachometer function, AC, 161-166

Tantalum capacitors, for high frequency bypassing, 43-44
TDA1180 horizontal-processor IC, 225
Television
  horizontal-sweep circuits in, 224-227
  sound channel IC for, 60-61
Temperature control, 235-238
Temperature tracking, transistor, 185
Three-phase power format
  from analog oscillator, 25-26
  from digital logic, 28-30
  from RC bridge network, 27-28
Three-terminal IC voltage regulators
  adjustable, 44-45
  and amplitude modulation, 222-224
  and ground connections, 46-47
  input voltage to, 48
  protection diodes for, 45-46
  regulation and ripple rejection by, 42
  stability in circuits using, 43
  thermal concerns with, 47-48
Threshold voltage, semiconductor, 116
Thyristors. *See* Diacs; GTOs; PUTs; SCRs
Tracking, negative voltage, 126-128
Tracking regulator, 109-111
Transients
  and MOVs, 167
  protection of digital-logic circuits from, 38-39
Transistors. *See* Bipolar transistors; Darlington transistors; MOSFETs
Transmitter, amplitude modulation, 219-221
Triacs
  400 Hz, 171
  driver for, optically coupled, 194, 197
  in electronic switch, 169
  light-controlled, 167-168
  load, table of, 206
  in motor phase control circuits, 149-151, 169-171
  in power interface, 38-39
  in remote switching, 194-197
  SCRs, compared to, 151
  in solid-state relay, 201-202
True power in polyphase power systems, 22
TTL logic. *See* Digital logic circuits
Twelve-volt amplifier systems, 68-71, 88-94
Two-phase power format, 30-32
Two-stereo-channel IC, 64-65
Two-terminal constant-current diodes, 107-108
UGS-3020 Hall-effect digital switch, 198-200
UJT (unijunction transistor) in motor control, 153-155
ULN-2280B audio IC, 62-64
ULN-2290B television sound IC, 60-61
ULN-3703Z/TDA2023 power amplifier IC, 68-69
Ultrasonic inverter, 100-watt, 227-228
Universal motors, control of, 13, 156-161, 169-171, 176
V13OLA2 MOV, 167

Voltage comparator, 118-119
Voltage drop
  bipolar transistor, 202
  junction diode, dealing with, 116
Voltage regulators
  using bipolar transistors, 105-107
  three-terminal. *See* Three-terminal IC voltage regulators
Voltage transients, and MOVs, 167

Voltage-frequency relationship in induction motor control, 179
Voltages, in polyphase power system, 20-28, 30
Wire, manganin, 120
Wirewound resistors in voltage regulators, 106
Wiring hints for three-terminal regulators, 49, 52
Y-connected loads in polyphase power systems, 21-22